Springer-AAS Acoustics Series

Series Editor

This series publishes peer reviewed high quality monographs and contributed volumes on all topics in acoustics. Books in this series range from those focused on a particular aspect of acoustics and vibration to practical handbooks covering a range of topics. The advantage for authors is that the inclusion of a book as part of this series will demonstrate high quality content of the book. While this series encourages authors and topics that are relevant to the Australasian region, proposals for contributions to the series are not be restricted to this region.

Rob Bullen

The Architecture of Sound

Acoustic Design For Buildings

Australian Acoustical Society

Springer

Rob Bullen
Rob Bullen Consulting
Neutral Bay, NSW, Australia

ISSN 2948-2062 ISSN 2948-2070 (electronic)
Springer-AAS Acoustics Series
ISBN 978-981-97-9379-2 ISBN 978-981-97-9380-8 (eBook)
https://doi.org/10.1007/978-981-97-9380-8

This Springer imprint is published by the registered company Springer Nature Singapore Pte Ltd.
The registered company address is: 152 Beach Road, #21-01/04 Gateway East, Singapore 189721, Singapore

Preface

This book is concerned with the design of sound within buildings. It aims to provide guidance on how to produce an appropriate acoustic environment for people using the building's spaces—that is, an environment that complements the use of the space, whether that is for work, relaxation, worship, enjoyment or any other use.

It will be of use to:

- those considering a career in acoustics as a specialist consultant,
- architects and design consultants with an interest in acoustics,
- anyone attempting to design or rectify the acoustic properties of building spaces (provided they already have some basic knowledge of acoustic principles) and
- lecturers designing a tertiary-level course for any of the above groups.

The book intends to fill a gap between popular non-mathematical characterisations of acoustic environments and highly technical descriptions that often appear divorced from real-world applications. It is organised in three parts.

Part I: Tools for Acoustic Design includes formulae and procedures for calculating sound levels in buildings, including the sound power propagated through a partition and along a duct. This section assumes competence with spreadsheet calculations using Microsoft Excel, as well as a basic understanding of the concepts of sound pressure level and sound power level. Spreadsheet files are included to automate many of the calculations.

Part II: Acoustic Design Principles introduces the four fundamental features of an acoustic environment that affect a user's experience:

- background sound,
- intruding sound,
- sound quality,
- speech intelligibility.

Each of these is covered in a separate chapter, with two chapters for intruding sound (airborne and structure-borne). In each case the chapter describes the effect of the feature in terms of human response; ways to characterise and measure it;

standards and guidelines for various spaces; and practical ways to achieve design requirements. Where guidelines are referenced, the focus is on those that apply in an Australian context, although others are also mentioned.

An additional chapter is included describing the challenges involved in implementing acoustic principles in conjunction with certain features of sustainable building design, and how those challenges can be met.

Part III: Design in Specific Spaces considers as examples two types of space—classrooms and hospital wards. In each of these we take a "deep dive" into how acoustic design principles operate in that setting, including information on the performance of typical spaces; research data on the effect of acoustic design on the functioning of the space; and examples of interventions that attempt to improve the acoustic environment.

Most chapters are accompanied by a downloadable spreadsheet containing data and calculation templates relevant to that chapter. In addition, most chapters contain exercises, which will generally involve calculations using a spreadsheet. Another downloadable file contains spreadsheets representing appropriate responses to those exercises.

Neutral Bay, Australia Rob Bullen

Acknowledgements The author is grateful for the advice and assistance of colleagues during the preparation of this manuscript, as well as the online course from which it has evolved. In particular, Glenn Leembruggen has provided valuable insight into speech intelligibility issues and prepared a draft of the relevant chapter. Marion Burgess has provided valuable advice on appropriate contents and materials for both the course from which this book has grown and for the book itself.

Contents

About the Author

Rob Bullen has over 40 years' experience in acoustic design and control, including over 20 published papers in the fields of environmental noise control, building acoustics and acoustic signal detection and separation. He has worked as lead acoustic consultant on projects including major airport developments, noise control for freeways and rail lines, and acoustic design for numerous building developments.

While working as a consultant, Rob has also developed a patented system for directional noise monitoring, produced publicly available software for the simulation of sound in internal environments and documented a system for separating a mixture of recorded sounds.

In addition, Rob teaches courses in acoustics for architects, designers and acoustic engineers at a number of universities. This book grew from teaching materials used in one such course, which is offered online by the University of NSW, Australia.

Part I
Tools for Acoustic Design

Chapter 1
Introduction

Abstract This Chapter provides a brief overview of the aims of the book and a map of the path to be followed in subsequent chapters.

Every building space has its own unique acoustic environment. In a conference room, the hush and the clarity of speech convey the message "every word is important". In a busy café, the jangle of voices, the whir of the coffee machine and sounds from the street all say "Dial up the energy. Time to get moving". In a library, the rattle of passing trains heard through underperforming glazing makes the space feel uncomfortable and somehow hostile. In a classroom, reverberation makes the room feel noisy, so that everyone speaks more loudly, creating more reverberation, eventually leading to disengagement by the students and stress for the teacher.

Whenever we create a space, we create an acoustic environment. This can be by accident or by design. The intention of this book is to show how an acoustic environment can be designed, so that the sound of the space complements its function. There are no simple rules for what makes "good acoustics", but there are principles that allow a designer to control the sounds that users will hear, their quality and their meaning.

There is no escaping the fact that good acoustic design involves calculations, and those calculations can be complicated and sometimes tedious. This book will not shy away from that fact. The first part, "Tools for Acoustic Design" sets out formulas and procedures required to calculate the basic descriptor of sound level—the sound pressure level spectrum—in various environments. The following part "Acoustic Design Principles" sets out the fundamental principles of acoustic design applicable to any space and in the process introduces more formulas for the calculation of descriptors such as reverberation time and speech transmission index. Familiarity with mathematical equations will be required in these sections.

To fully understand the design process, it will be useful to attempt the many exercises scattered throughout the book. Most chapters include a downloadable Excel spreadsheet which provides data and/or calculation templates to carry out some of the more complex calculations described in the chapter. A separate spreadsheet demonstrates appropriate responses to the exercises in the chapter.

The final part of the book "Design in Specific Spaces" shows how the general acoustic design principles described in Part 2 apply to two specific types of space—classrooms and hospital wards. These serve as examples of the practical acoustic issues that can arise in particular spaces and of the difference that good acoustic design can make to the way that they function. They serve as examples of how to apply general principles of acoustic design to specific types of space or to particular individual spaces.

R. Bullen, *The Architecture of Sound*, Springer-AAS Acoustics Series,
https://doi.org/10.1007/978-981-97-9380-8_1

This book is intended to provide practical advice for people who will be responsible for designing acoustic environments. It is not intended to provide a full, thorough discussion of the acoustic and psychoacoustic processes involved in human response to sound. The current state of knowledge in this area is covered in other textbooks, two of the most significant being by Pierce (2019) and Everest (2014). These are recommended for readers interested in a more in-depth (and more mathematical) coverage of these issues.

References

Everest A (2014) Master handbook of acoustics. McGrawq Hill Education Tab
Pierce A (2019) Acoustics: an introduction to its physical principles and applications. Springer Cham

Chapter 2
Sound Pressure Level in a Room

Abstract This chapter first ensures that readers understand the basic requirements for spreadsheet calculations and the concepts of sound level, sound power and directionality that will be used throughout the book. It then introduces "Sabine" and "Schultz" rooms, and shows how to calculate the sound pressure level (SPL) in each of these rooms due to a source with a known sound power level (SWL). It also indicates how to calculate SWL from a measured value of SPL.

2.1 Sound Pressure Level and Sound Power Level

Fundamental to the description of any sound is the sound pressure level (SPL), which represents the mean square sound pressure measured at some location, expressed as decibels relative to the base pressure of 2×10^{-5} Pa. That is,

$$\mathrm{SPL} = 10\,\lg\left(\frac{\langle p^2\rangle}{p_0^2}\right) \tag{2.1}$$

where p represents sound pressure, p_0=2×10^{-5} Pa and $<>$ represents a time average. If the sound pressure is filtered to include only components within a particular octave band or third-octave band of frequencies, then *SPL* will be the octave or third-octave band sound pressure level. A set of octave or third-octave band SPL values covering the range of frequencies in the sound is called an octave or third-octave band *spectrum*.

SPL is directly related to the subjective loudness of the sound, although this relationship is complex and mediated by factors related to the frequency content, level and temporal characteristics of the sound. A detailed description of these factors can be found in Moore (2013). A simple and very common way to approximate loudness is by the A-weighted value of the SPL, which can be found by either:

Supplementary Information The online version contains supplementary material available at https://doi.org/10.1007/978-981-97-9380-8_2.

R. Bullen, *The Architecture of Sound*, Springer-AAS Acoustics Series,
https://doi.org/10.1007/978-981-97-9380-8_2

- directly filtering the sound using a filter with certain defined specifications (as set out in Acoustical Society of America (2023) and implemented in complying sound level meters) or
- taking the spectrum of the sound, applying defined "A-weighting" corrections to each octave or third-octave band, and adding the results logarithmically. This process is approximate and is more accurate if a third-octave spectrum is used.

It is assumed that the reader has some familiarity with these concepts—if not they are advised to consult any introductory text on acoustics such as (Randall 2005). (A more detailed discussion than is presented here can be found in texts such as (Pierce 2019)).

Many of the calculation procedures described in this book, including the exercises, involve manipulations carried out on sound spectra. These are most easily carried out in a spreadsheet program. Appendix 1 provides a briefing on the use of Microsoft Excel for calculations on spectra, including how to calculate the A-weighted level.

Exercise 2.1
Take the SPL from a normal voice at 1 m directly in front, from the "Sound Levels from Voice" data sheet in Supplementary file1 spreadsheet. Using a format similar to the "First Example Calculation" data sheet:

- add a correction of +6 dB at all frequencies,
- then logarithmically add the SPL from a raised voice at the same position.

Find the A-weighted level of the total.

The sound pressure level produced by a sound source depends on both the source and the measurement position—in general SPL will decrease with increasing distance from a source. In contrast, *sound power* is a property of the source only and represents the total sound energy emitted by the source per second, in all directions. It is expressed in decibel terms by the sound power level (SWL), defined as

$$\mathrm{SWL} = 10\,\mathrm{lg}\left(\frac{W}{W_0}\right) \tag{2.2}$$

where W represents the sound power of the source in Watts and $W_0 = 10^{-12}$ W.

The sound sources referred to in this book may be *point sources*, in which case they are small compared with a typical measurement distance. Examples include a person talking, or an air-conditioning outlet. Alternatively, they may be *surface sources* that represent an extended area such as a wall or a window. These types both have an associated sound power. Surface sources are sometimes characterised by the sound power emitted per m^2, but in this book the term SWL will always represent the total sound power level emitted by the source.

2.2 Relationship Between SPL and SWL—Point Source in Space

Much of acoustics is concerned with calculating the SPL due to sources with known SWL when they are placed in various environments. The very simplest case considers an omnidirectional source (i.e. one emitting equal sound power in all directions) in a "free field" (i.e. with no nearby surfaces, including the ground). In this case, the relationship between SPL and SWL is

$$\mathrm{SPL} = \mathrm{SWL} + 10\lg\left(\frac{1}{4\pi r^2}\right)$$

or

$$\mathrm{SPL} = \mathrm{SWL} - 20\lg(r) - 11 \tag{2.3}$$

Equation (2.3) is often inverted to find the SWL of a source from a measured SPL:

$$\mathrm{SWL} = \mathrm{SPL} + 20\lg(r) + 11 \tag{2.4}$$

If the source is directional, the measured SPL will depend on the direction in which it is measured. Measurements are usually in the horizontal plane, and the direction of the measurement is given by a single angle θ. The *directionality at angle* θ, $Q(\theta)$, is defined so that the SPL measured at angle θ is

$$\mathrm{SPL}(\theta) = \mathrm{SWL} + 10\ \lg\left(\frac{Q(\theta)}{4\pi r^2}\right)$$

or

$$\mathrm{SPL}(\theta) = \mathrm{SWL} - 20\ \lg(r) - 11 + 10\ \lg(Q(\theta)) \tag{2.5}$$

For an omnidirectional source, $Q(\theta) = 1$ for all values of θ, so $10\lg(Q(\theta)) = 0$. For a directional source, the average value of $Q(\theta)$ over all angles is always 1.

For a directional source, Eq. (2.4) is still valid for calculating *SWL* from *SPL provided* that *SPL* in the equation is the *logarithmic average* SPL calculated over all values of θ.

When the measurement is conducted over a hard ground (or floor), "free-field" measurement conditions may still apply depending on the measurement distance. Figure 2.1 shows a source and receiver, assumed to be at the same height above a hard ground. If the measurement distance r is less than the measurement height h, the measurement conditions can be assumed to be free field. If r is between about 3 h and 7 h, Eqs. (2.3), (2.4) and (2.5) can apply with the constant 11 replaced by 8. This is sometimes called the "hard ground" case.

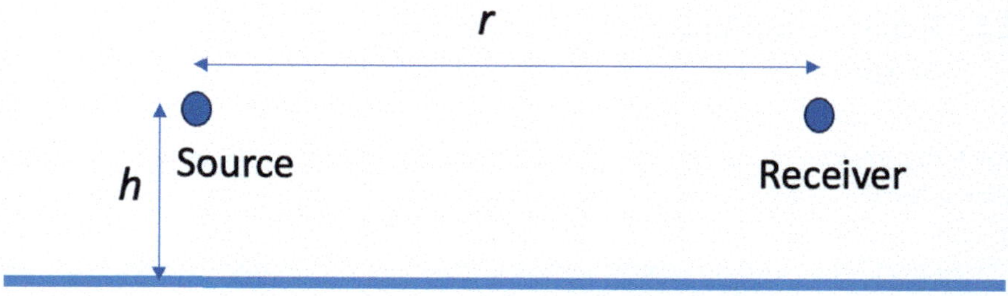

Fig. 2.1 Source and receiver locations above a hard ground

Otherwise, more complex calculation procedures need to be used. These are described in more mathematically based texts such as Pierce (2019).

2.3 Time-Varying Sound

When sound levels vary with time, the perceived loudness varies in complex ways that depend on the properties of the sound spectrum, the pattern of temporal variation and other factors. (See Moore (2013) for a full discussion of these factors.) For practical measurements, an approximation to the loudness at any time is required.

The time-varying sound pressure level is typically characterised by an average of the squared sound pressure with a weighting function prioritising times within a small window of the measurement time. This is expressed mathematically as

$$\mathrm{SPL}(t) = 10\lg\left(\frac{1}{\alpha}\int_{-\infty}^{t}\frac{p^2(t)}{p_0^2}e^{-(t-\tau)/\alpha}d\tau\right) \tag{2.6}$$

in which α represents a "time constant" that determines the width of the time window. In typical measurements, one of two standard constants is used: $\alpha = 125$ ms for "Fast" speed measurements or $\alpha = 1$ s for "Slow" speed measurements.

An important property of sounds considered in building acoustics is the maximum loudness achieved, which is typically approximated by the maximum value of the A-weighted SPL measured using the fast speed window. This is written $L_{AF\,\mathrm{max}}$.

An alternative descriptor is based on the average of the squared sound pressure taken over a long time period—effectively just Eq. (2.1) where the averaging time is measured in minutes, or even hours. This descriptor is written L_{eq}, or $L_{A\mathrm{eq}}$ if the sound pressure is A weighted. It can alternatively be described as the logarithmic average of SPL(t) values over the time period in question. (This average is independent of the time constant used.)

In an acoustic environment, L_{eq} values of different component sounds can be added logarithmically to give the L_{eq} of the total sound. This is of course not true of $L_{AF\,\mathrm{max}}$ values—these would add logarithmically only if the maximum values of all component sounds happened at exactly the same time.

2.4 Using the Supplied Voice Data

As examples of SPL, directivity and SWL data, the third sheet of the supplied Supplementary file1 spreadsheet contains the following data for the human voice (averaged over male and female voices):

- SPL spectra for normal voice, raised voice, loud voice and shout, at 1 m directly in front of the talker,
- values of $10\lg(Q(\theta))$ by frequency, for angles between 0 and 180 degrees from the front of the talker and
- SWL spectra for each voice level, calculated from the logarithmic average SPL over all angles using Eq. (2.4).

These data are taken from (Acoustical Society of America 1997). Note that the values are L_{eq} for each frequency band—that is, they represent time-averaged sound levels over a typical section of speech. $L_{AF\,\max}$ sound levels would be approximately 5 dB higher than the L_{Aeq} value

Exercise 2.2
Find the L_{eq} spectrum of a raised voice heard at an angle of 120 degrees from the front, at 5 m in a free field.

2.5 Relationship Between SPL and SWL in a Room

A prerequisite for designing an acoustic environment is to understand how to calculate the SPL due to a source at any point in the environment.

This section applies to *rooms*, which are spaces that are fully or almost fully enclosed, and then only to rooms that can be approximated by one of two "standard" types. For non-enclosed spaces or unusual room configurations, calculation of the SPL is more complex and may require a computer model as described in Appendix 3.

The two "standard" room types are described below, and in more detail in Appendix 2. Appendix 3 explains the basis for the formulas relating SPL to SWL in these rooms. This chapter simply states the formulas. These will be needed in many places in subsequent chapters throughout this book.

2.5.1 "Sabine" Rooms

These are rooms in which there is significant reverberation and/or significant sound diffusion, which creates a "reverberant" sound field that fills the entire room. Close to the source, there is also a "direct" field due to sound reaching the listener without being reflected from any room surface. Good examples of Sabine rooms are mechanical plant rooms and concert halls.

In these rooms, the SPL anywhere in the room due to a *point* source is given by

$$\mathrm{SPL}_R = \mathrm{SWL} + 10\,\lg(T) - 10\,\lg(V) + 14 \tag{2.7}$$

$$\mathrm{SPL}_D = \mathrm{SWL} + 10\,\lg(Q(\theta)) - 20\,\lg(r) - 11 \tag{2.8}$$

$$\mathrm{SPL} = 10\,\lg\left(10^{\mathrm{SPL}_R/10} + 10^{\mathrm{SPL}_D/10}\right) \tag{2.9}$$

in which

- SPL_R and SPL_D are respectively the contributions from the reverberant and direct sound fields.
- SWL is the sound power level of the source.
- T is the reverberation time of the room in seconds (described in more detail in Chap. 8).
- V is the volume of the room in m^3.
- $Q(\theta)$ is the directivity of the source in the direction of the listener, as described in Sect. 2.2.
- r is the distance from source to listener in m.

Note that Eq. (2.8) is simply Eq. (2.5)—the "direct" field is the same as a free field.

In general, SWL, T and Q all depend on frequency, so calculation of SPL requires a spreadsheet as described in Appendix 1.

For a *surface* such as a wall in a Sabine room, the contribution of SPL_D will be negligible, and the total SPL is just given by Eq. (2.7).

2.5.2 "Schultz" Rooms

Many "normal" rooms such as domestic spaces, offices and cafes do not meet the requirements of a Sabine room, but still exhibit regularities in the relationship between SPL and SWL. A paper by Schultz (1985) postulates a general relationship valid for these spaces, namely

$$\mathrm{SPL} = \mathrm{SWL} - 3\,\lg(f) - 5\,\lg(V) - 10\,\lg(r) + 12 \tag{2.10}$$

where V and r are as above, and f is an octave-band or third-octave band centre frequency in Hz. Notice that in these spaces there is no reverberant field that is constant everywhere.

The sources considered by Schultz were approximately omnidirectional. For a directional source, it is reasonable to simply replace SWL in (2.10) with SWL + $10\lg(Q(\theta))$ where θ is the direction from source to listener.

Equation (2.10) is valid for a *point* source in a Schultz room. For a surface like a wall or window, we need an additional correction, which is described in Appendix 3. If r now represents the *shortest* distance to the surface, and the surface dimensions are $L_1 \times L_2$ with L_1 being the longer side, then

$$\begin{aligned} \text{SPL} = \text{SWL} &+ 10\lg(Q(\theta)) - 3\lg(f) - 5\lg(V) \\ &- 10\lg(r) + 12 + \Delta_s \end{aligned} \tag{2.11}$$

where

$$\Delta_s = 10\lg\left\{\frac{2}{\overline{\rho}^2}\left[\sqrt{1+\overline{\rho}^2} - 1\right]\right\}$$

and

$$\overline{\rho}^2 = \frac{L_1\sqrt{L_1 L_2}}{\pi r^2}$$

For a point source, $\Delta_s = 0$.

2.6 Using the Supplied Calculation Spreadsheets

To avoid having to type in the above formulas whenever they are needed, the calculations can be performed in the fourth and fifth sheets of the supplied Supplementary file1 spreadsheet. These sheets calculate the *difference* between SPL and SWL for Sabine and Schultz rooms, for point or surface sources. The calculated differences can be added to the SWL of the source to give the SPL in the room.

Using these sheets involves entering the relevant room parameters and copying the results to a separate calculation sheet. (When the values to be copied are calculated, make sure that you paste them using *Paste Special > Values*.)

As in all calculation spreadsheets supplied with this book:

- green cells indicate locations where input is expected,
- orange cells indicate calculated output,
- grey cells indicate intermediate calculations.

The calculations in Supplementary file1 spreadsheet are in third-octave bands. If values in octave bands are needed, just use the values at the octave band centre frequencies.

Exercise 2.3
Using SWL and directionality data in Supplementary file1 spreadsheet, calculate the SPL in dBA from a person speaking in a normal voice at a distance of 3 m in:

(a) a Sabine room with a volume of 150 m^3 and reverberation time 0.8s at all frequencies, directly in front of the talker
(b) a Schultz room with a volume of 150 m^3, directly in front of the talker

2.7 Finding the SWL of a Source

How do we know the SWL of a sound source? There are at least five possibilities.

1. The SWL is specified by the manufacturer. This applies mostly to mechanical items including fans, motors, pumps, etc. Measurements can be assumed to have been made in accordance with one of a number of relevant international Standards, which are described in (International Standards Organisation 2019). The manufacturer will supply the SWL in octave or (rarely) third-octave bands. Often the SWL will vary depending on operating conditions and you will need to know what conditions apply in your case.
2. The SPL has been measured at a known distance in a free field, so the SWL can be calculated from Eq. (2.4), as for the speech data set in Supplementary file1 spreadsheet.
3. The SWL is set by a required A-weighted SPL in a room. An example is a television. The listener will want a certain SPL at their listening position (say, L_{Aeq} of 65 dBA from normal speech) and will adjust the audio volume until that is achieved. In this case, to find the SWL produced by the television one needs to work backwards using the equations above to find what SWL will give the required SPL in that room, at that listening position.

   The easiest way to do that is:

   - start with a relative spectrum for the sound—that is, make the shape of the spectrum correct but don't worry about the absolute values,
   - find the total SPL at the listening position with that spectrum as the SWL, using the relevant calculation sheet in Supplementary file1 spreadsheet,
   - then adjust the original spectrum at each frequency by the difference between the required A-weighted SPL and the calculated A-weighted SPL. This will give an SWL spectrum that will produce the required SPL at the listening position.

4. The SWL enters the room from another space, through a partition. This case is described in Chap. 3.
5. The SWL enters the room through a duct, generally as part of an air-conditioning system. This case is described in Chap. 4.

In all cases, be careful to note whether the SWL values you are using represent L_{eq} or $L_{AF\max}$. For example, for speech, in some circumstances (e.g. speech intelligibility assessment) the L_{eq} value is required, while in others (e.g. noise intrusion assessment) it is the $L_{AF\max}$.

Exercise 2.4
An omnidirectional point source with the spectrum of a raised voice produces an SPL of 70 dBA L_{Aeq} at 2 m in a Schultz room with volume 60 m^3.

(a) What is the source's A-weighted SWL?
(b) What would be the A-weighted SPL from this source at 5 m in the same room?

Appendix 1: Spreadsheet Manipulations

Spreadsheet Basics

When following this book it will be useful to try the calculation exercises which are scattered throughout. Many exercises require the use a spreadsheet, and the industry standard programme for this purpose is Microsoft Excel. The downloadable spreadsheets accompanying this book are formatted in Excel and may not function properly in other programmes.

Calculations will involve additions, subtractions, and logarithmic additions over a range of octave-band or third-octave band centre frequencies, together with logarithmic addition across frequencies. Calculations can be set out with frequencies running either horizontally or vertically. Most examples used in this book will have frequencies set out horizontally, but some calculation spreadsheets have them vertically.

A calculation spreadsheet using octave-band frequencies, set out horizontally, might look like Fig. 2.2. This is a copy of the second sheet of the Excel spreadsheet Supplementary file1.xlsx which is supplied for this Chapter, and it is worthwhile to open that example and understand it.

The bands used for a calculation may be different from those shown, and they may be third-octave bands rather than octave bands. Decibel values are usually shown to one decimal place in calculations. However, the final result will often be quoted to the nearest decibel.

	EXAMPLE CALCULATION							
Frequencies	**63**	**125**	**250**	**500**	**1000**	**2000**	**4000**	**A-weighted**
A-weighting correction	**-26.2**	**-16.1**	**-8.6**	**-3.2**	**0.0**	**1.2**	**1.0**	
Begin with this spectrum	60.0	65.0	68.0	70.0	65.0	63.0	55.0	**70.8**
A correction may be required in which values are added or subtracted								
If these values are subtracted:	5.0	6.0	7.0	8.0	9.0	10.0	11.0	
We get this spectum	55.0	59.0	61.0	62.0	56.0	53.0	44.0	**62.2**
Another spectrum may need to be added logarithmically								
If this spectrum is added logarithmically:	55.0	55.0	53.0	51.0	50.0	59.0	48.0	
We get this spectum	58.0	60.5	61.6	62.3	57.0	60.0	49.5	**64.7**

Fig. 2.2 Example of a calculation spreadsheet

Logarithmic Addition

To logarithmically add decibel values in the cells B7 and B8 (for example), enter the formula

$$= 10^{*}\mathrm{LOG10}\left(10^{\wedge}(\mathrm{B7}/10) + 10^{\wedge}(\mathrm{B8}/10)\right)$$

Alternatively this could be

$$= 10^{*}\mathrm{LOG10}\left(\mathrm{SUM}\left(10^{\wedge}((\mathrm{B7} : \mathrm{B8})/10)\right)\right)$$

In Excel, LOG10 can alternatively be written as simply LOG. Writing LOG10 just makes the formula slightly clearer.

To logarithmically *average* decibel values in cells B7:B10, use the formula

$$= 10^{*}\mathrm{LOG10}\left(\mathrm{AVERAGE}\left(10^{\wedge}((\mathrm{B7} : \mathrm{B10})/10)\right)\right)$$

A Weighting

The best way to find the A-weighted total of a spectrum is to use a feature of Excel called an *array formula.* To demonstrate this, type a formula that involves operations on two ranges range of cells—for example

$$= \mathrm{SUM}(\mathrm{B5} : \mathrm{H5} * \mathrm{B6} : \mathrm{H6})$$

and after typing, instead of pushing Return, push Ctrl-Shift-Return. The formula will then have braces around it like this { }, and the cell now shows the value of

$$B5 * B6 + C5 * C6 + \ldots$$

To use this feature, the appropriate A-weighting corrections should be entered below each frequency band as in Fig. 2.2. If these are in cells B3:H3 and a spectrum is in B5:H5, the total A-weighted value of the spectrum will be

$$= 10\text{*}\mathrm{LOG10}(\mathrm{SUM}(10\text{^}((B3 : H3 + B5 : H5)/10)))$$

provided you push Ctrl-Shift-Return after entering the formula.

To copy this to another spectrum, make the A-weighting corrections constant:

$$= 10\text{*}\mathrm{LOG10}(\mathrm{SUM}(10\text{^}((B\$3 : H\$3 + B5 : H5)/10)))$$

You can now copy and paste the formula to another line and get the A-weighted value of the spectrum in that line.

Appendix 2: The Impulse Response

Impulse Response of a Space

The impulse response of a room can be thought of as the sound pressure produced at a receiver position due to a very brief sound (such as a balloon burst) at a source position. (Mathematically, it represents the response to a pulse of pressure occurring over a vanishingly small time.)

The impulse response is a pressure-versus-time waveform, recorded at audio frequency. It can be either:

- scaled, in which case the pressure values are relative to the pressure resulting from the source at some reference distance in an anechoic environment,
- unscaled, in which case the absolute pressure values are typically chosen to optimise recording quality.

Calculated impulse responses are generally scaled.

If the impulse response is known, it can be used to calculate any feature of the acoustic environment at the receiver location due to a source at the source location. This includes:

- the reverberation time, as a function of frequency if required,
- an audio simulation of the sound at the receiver due to any sound produced at the source,
- if scaled, the SPL due to a source of known SWL.

The impulse response can be estimated by recording the response to a sound such a balloon burst. However, more accurate methods exist which involve

- producing one of a range of specific sounds—often a swept-frequency tone—at the source via a loudspeaker,
- recording the response at the receiver,
- mathematically processing the result.

These techniques produce a significant increase in signal-to-noise ratio, allowing the sound decay to be followed over a longer time period.

To achieve a scaled response, the source must be relocated and measured either in an anechoic environment or at a location where direct sound can be clearly separated in time from any reflections.

Current measurement and processing equipment allows impulse responses to be measured relatively easily and quickly, and measurement of the impulse response of a space is a popular analysis technique.

Figure 2.3 shows an example of an impulse response measured in a lecture room at approximately 4 m from the source. (This is taken from the dataset described in Section 2.2.3.)

If the impulse response is squared and integrated over short time windows, it represents the decay in SPL after the impulse is produced. This is shown in Fig. 2.4, using a time window of 10 ms. The response is characterised by:

- a pulse representing the arrival of direct sound from the source,
- a period of fast decay, including other early reflections, before true reverberation from the room becomes significant,
- a period of approximately exponential decay, which is linear when expressed as decibels, representing room reverberation.

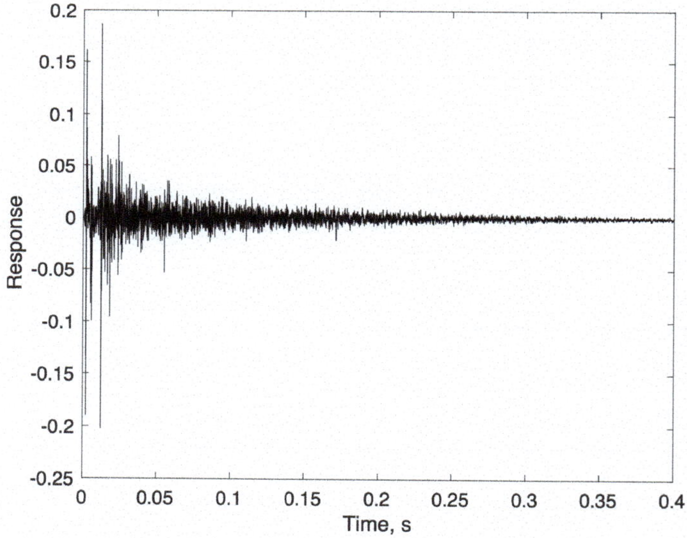

Fig. 2.3 Impulse response recorded in a lecture room at approximately 4 m from the source

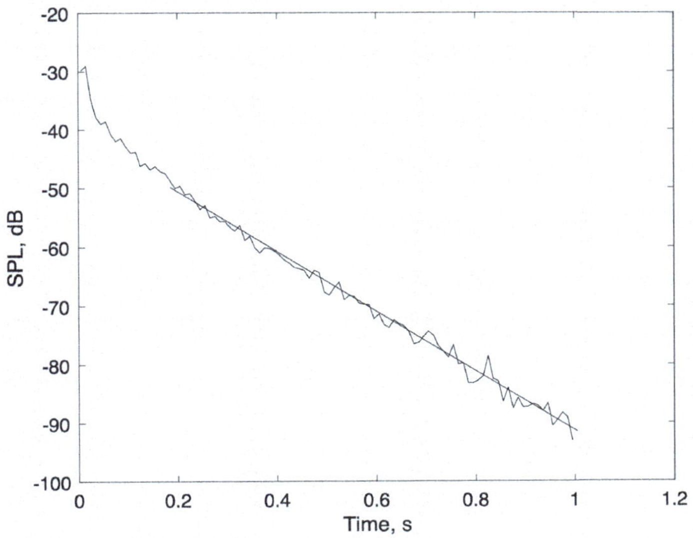

Fig. 2.4 Impulse response in Fig. 2.3 converted to (relative) SPL versus time—time window 10 ms

Figure 2.5 shows the same impulse response integrated with a time window of 1 ms. Here, the influence of individual reflections can be more clearly seen.

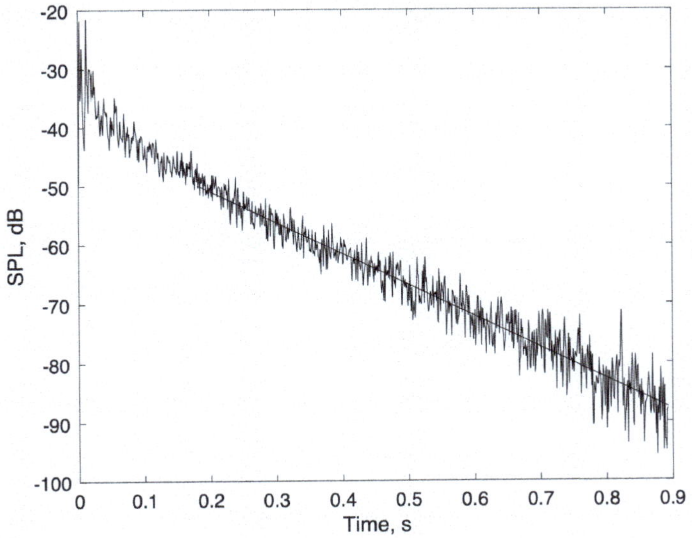

Fig. 2.5 Impulse response in Fig. 2.3 converted to (relative) SPL versus time—time window 1 ms

Calculating the Impulse Response

The impulse response of a space arises from the addition of many "image sources", each generated by reflection of the original impulse from multiple surfaces in the space, like reflection of light from multiple mirrors. As the number of reflections increases, the image sources become "smeared" copies of the original, due to absorption and diffusion in the reflecting surfaces as well as diffraction of sound around obstacles.

If a detailed 3D model of the space is available, this can be entered into any of several computer modelling programmes that will trace each reflection and produce a calculated impulse response. The programmes generally require significant time to set up and are not exact—for example, many ignore the "smearing" of image sources described above. However, the results generally give an acceptable match to measured impulse responses.

If detailed modelling is not possible, then statistical methods can be used to estimate the SPL at times not too close to the original impulse.

As time increases, so does the number of image sources arriving at the receiver position per unit time. By calculating

- the average number of image sources arriving per unit time,
- the average number of reflections experienced by each source,
- the average reduction in sound power due to those reflections (and assuming they occur from random surfaces in the space),

it is possible to estimate the sound pressure level at the receiver as a function of time. (See (Hopkins 2007) for a detailed description of this calculation.) The result is that the SPL is predicted to reduce linearly, with a slope of

$$\frac{10c}{\ln(10)}\frac{S}{4V}\ln(1-\alpha)$$

per second, in which c is the speed of sound, S and V are the total surface area and volume of the space and α is the mean absorption coefficient of the space. That is,

$$\alpha = \frac{\sum_i S_i\alpha_i}{S}$$

where S_i and α_i are respectively the surface area and absorption coefficient of each surface in the space. Taking the reverberation time RT as the time for the SPL to reduce by 60 dB, this leads to

$$RT = \frac{24\ln(10)}{c}\frac{V}{S(-\ln(1-\alpha))}$$

$$\sim .161\tfrac{V}{S(-\ln(1-\alpha))} \tag{2.12}$$

if all quantities are in SI units, which is known as the Norris-Eyring reverberation time. If α is small, this can be approximated by

$$RT \sim .161\frac{V}{S\alpha} \tag{2.13}$$

which is known as the Sabine reverberation time. This is the most commonly-used formula for calculating the reverberation time of a space.

Note that the decay of SPL will be linear only when image-source contributions are sufficiently closely spaced that their contributions can be averaged. For example, in Fig. 2.4, this would be after approximately 0.2s. The time required depends on a number of factors, including:

- the shape of the space. In spaces in which one dimension is very different from the others (e.g. corridors, or large rooms with low ceiling height), the time required before the decay becomes linear is increased. In some open-ended spaces such as tunnels or courtyards, linearity may never be achieved,
- diffusion of surfaces. Where surfaces produce more diffuse reflections, the time to reach linearity is reduced,
- placement of absorption. The time to reach linearity is reduced if absorption is distributed relatively evenly among the surfaces in the space,
- frequency dependence of absorption. If certain frequencies are absorbed much more strongly than others, the decay of (unweighted) SPL vs. time will not be linear.

When conditions are such that image-source contributions can be averaged and the decay of SPL with time is linear, the sound field is said to be "diffuse".

A space where diffuse conditions apply from very early in the decay is sometimes referred to as a "Sabine" space, and that terminology is adopted in this book. In a Sabine space, the decay curve will be linear for almost its entire length and its slope will be described by the Sabine reverberation time. Such spaces are generally quite reverberant and include plant rooms and "reverberation rooms" designed for acoustic testing.

Examples of Integrated Impulse Responses

Szoke et al. (2019) describe a study which involved recording numerous unscaled impulse responses in a range of spaces. The responses are publicly available at https://speech.fit.vutbr.cz/software/but-speech-fit-reverb-database

Responses were selected from seven spaces in this set, with an additional response recorded in a concert hall, as described in Merimaa et al. (2005) and available at http://legacy.spa.aalto.fi/projects/poririrs/

Figure 2.6 shows the integrated impulse responses recorded in this selection of spaces. These responses are measured at various distances from the source (see

Table 2.1). Different locations in the same room will produce different responses. These responses have no frequency weighting.

Impulse Response and Reverberation

Reverberation of sound in an enclosed space is described in Chap. 8, where it is characterised by the reverberation time. This section describes the connection between reverberation time and the impulse response.

Although reverberation time is often loosely defined as "the time taken for the sound level to decay by 60dB", Australian Standard 2460:2002 "Acoustics—Measurement of reverberation time in rooms" (Standards Australia; Standards New Zealand 2002) makes it clear that under its definition, the reverberation time is based on the slope of the decay curve *in the region where that slope is linear*. As seen in Fig. 2.6, the linear section of the decay may begin very soon after the arrival of the direct sound, as for "Shared Office", or may be delayed, as for "Meeting Room".

In this book, "reverberation time" or RT (with no subscript) will be taken to be the time as defined in AS 2460. This will generally be based on the slope of the curve later in the decay (but before the trace is influenced by ambient noise) as in Fig. 2.6. In some traces, there may be no linear section, in which case RT is undefined.

Other definitions of reverberation time, notably as found in International Standards Organisation (2008a), are described below.

AS 2460 and other sources describe two methods of measuring reverberation in a room—the "interrupted noise source" and "impulse noise source" methods. The "impulse noise source" method gives the impulse response directly (subject to measurement restrictions). In contemporary practice, the "impulse" method is generally preferred as it provides information on the room's acoustic characteristics in addition to the reverberation time.

Results from the "interrupted noise source" method can be calculated from the "impulse" results, but not vice-versa. This calculation can be performed in one of two ways:

- A digital signal can be generated representing the interrupted noise, and this can be convolved with the impulse response. (Convolution is a mathematical procedure described in any textbook on digital signal processing.)
- A simpler method involves "reverse integration" of the impulse response using a formula developed by Schroeder and described in International Standards Organisation (2008b).

In either case, care must be taken to truncate the recorded impulse response before it is significantly affected by background noise.

Reverberation traces calculated in this way confirm that when the decay is linear, the slope of the decay is very similar for the two measurement methods (Fig. 2.7). However if the decay is substantially nonlinear, the two methods produce significantly different decay traces (Fig. 2.8).

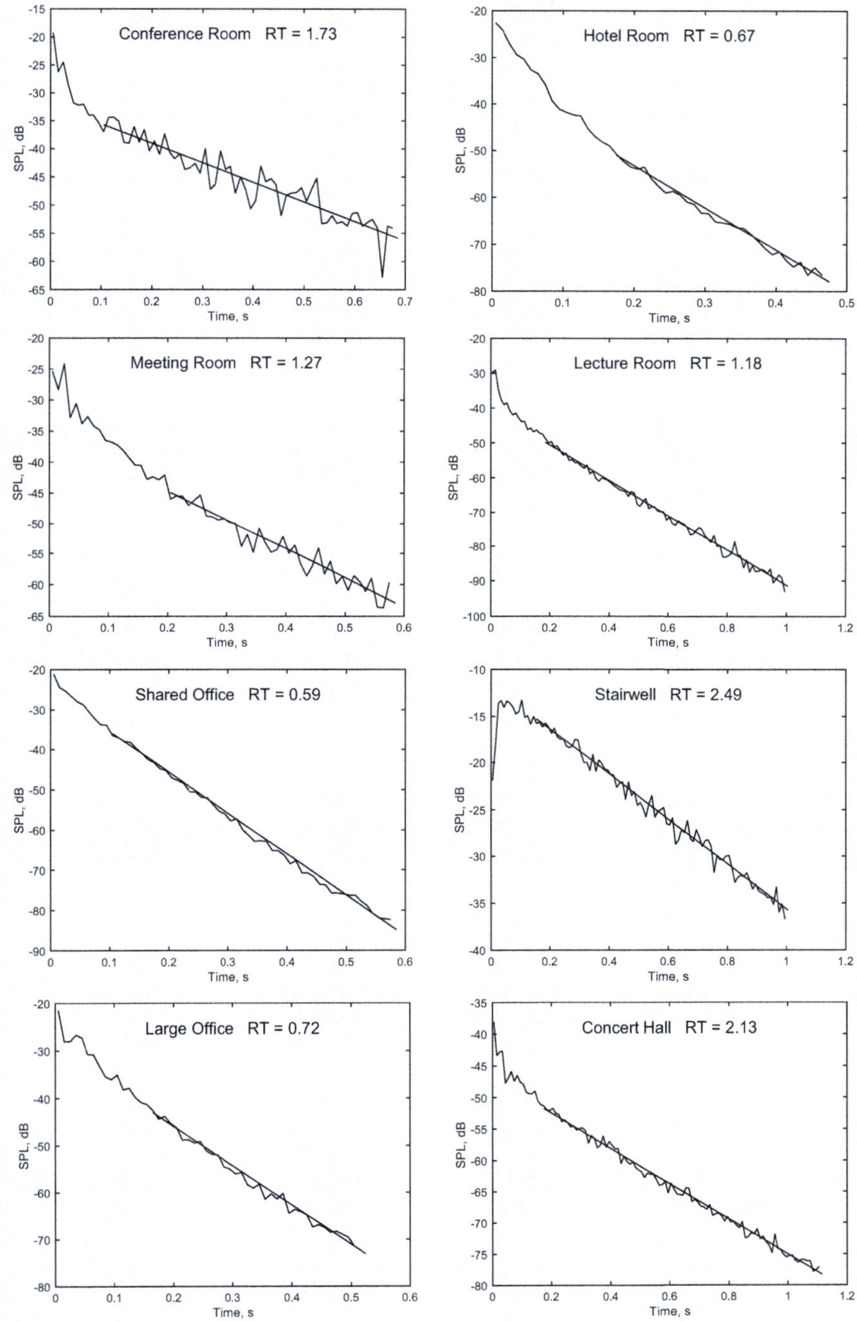

Fig. 2.6 Reverberation traces in typical environments, calculated from data described in (Szoke et al. 2019) and (Merimaa et al. 2005)

Table 2.1 Properties of the example spaces (Open the associated Supplementary file to hear speech as heard in the space)

Space	Distance, m	EDT, s	T_{20}, s	T_{30}, s	RT, s	C_{50}, dB
Conference room [Link to Supplementary file3}	3.9	0.16	1.22	1.33	1.73	5.2
Hotel room [Link to Supplementary file4}	2.0	0.27	0.37	0.42	0.67	8.8
Meeting room [Link to Supplementary file5}	2.8	0.27	0.75	1.09	1.27	4.7
Lecture room [Link to Supplementary file6}	4.2	0.22	0.71	0.91	1.18	6.1
Shared office [Link to Supplementary file7}	2.1	0.35	0.51	0.53	0.59	5.7
Stairwell [Link to Supplementary file8}	4.1	2.26	2.21	1.48	2.49	-7.7
Large office [Link to Supplementary file9}	3.8	0.35	0.58	0.61	0.72	5.7
Concert hall [Link to Supplementary file10}	9.8	0.58	1.62	1.89	2.13	0.4

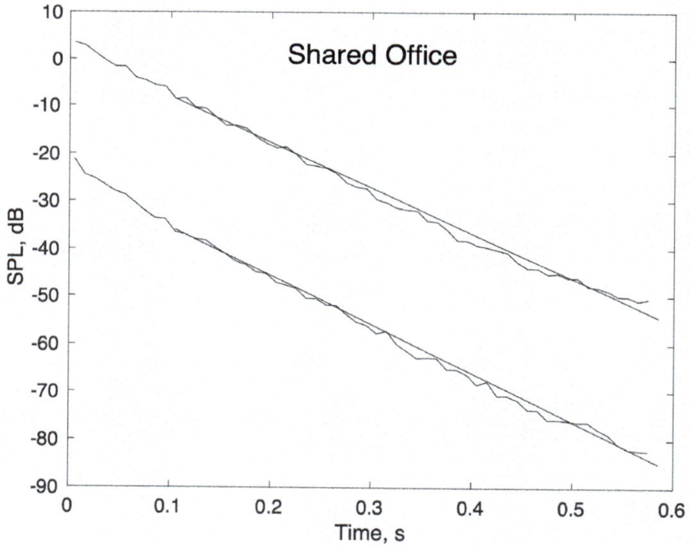

Fig. 2.7 Reverberation traces for the shared office calculated using the "impulse" method (lower) and "interrupted noise source" method (upper)

Of course, complete description of the acoustic characteristics of a space will require measurements for multiple source and receiver positions. Each combination of source and receiver will have a different impulse response. The number of measurements required will depend on the complexity of the space and the purpose of the measurements. To characterise the properties of an "ordinary room", ISO 3382

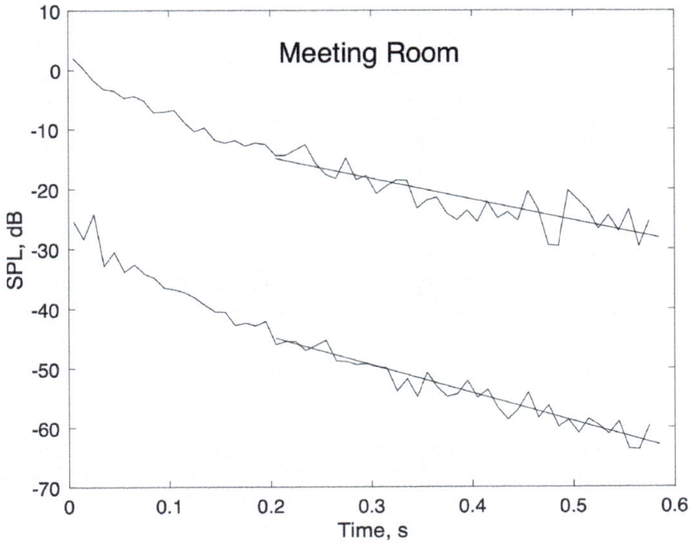

Fig. 2.8 Reverberation traces for the meeting room calculated using the "impulse" method (lower) and "interrupted noise source" method (upper)

recommends that the minimum number of source-microphone combinations should be:

- 2 for "survey" measurements,
- 6 for "engineering" measurements,
- 12 for "precision" measurements.

Reverberation times can be arithmetically averaged to give an overall value for the space.

It should be noted that the recorded impulse response will depend on the directivity of the sound source (generally a loudspeaker). AS 2460 recommends that this should be as omnidirectional as possible.

Frequency Dependence of the Reverberation Time

Reverberation calculated directly from an impulse response, as above, represents the room's response to a white noise signal. However, the reverberation time in any frequency band can be found either by filtering the impulse response or by using the "interrupted noise source" method with a signal consisting of filtered noise.

Figure 2.9 shows octave-band reverberation times for the example spaces described above. Points of interest include the following.

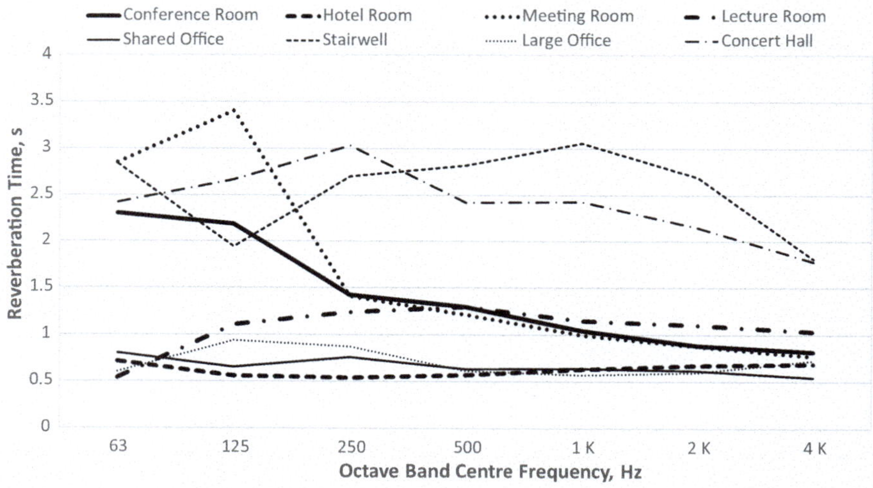

Fig. 2.9 Reverberation time by octave-band centre frequency for spaces in Fig. 2.6

- The conference room and the meeting room both have significant differences between low-frequency and high-frequency RTs (at least, at the sampled source and receiver locations). This explains their nonlinear reverberation traces for a mixture of frequencies.
- The conference room, lecture theatre and meeting room all have mid-frequency RTs well above recommendations (see main text, Sect. 2.5).
- The stairwell and concert hall have similar RTs, but sound very different (see below).

Other "Reverberation Time" Descriptors

As seen in Fig. 2.6, if the reverberation curve is not linear, then reverberation time, or RT as defined here, refers to the slope of the curve in the late stage of the decay, with faster decay generally being evident at earlier times. RT describes the "noisy" section of the decay, while the earlier sections contain information from the source.

The most popular metrics to describe the earlier section of a decay are:

- RT_{20}, based on the slope of the decay between −5 dB and −25 dB relative to the maximum level,
- RT_{30}, similarly defined between −5 dB and −35 dB,
- Early Decay Time (EDT) based on the slope between the maximum level and the maximum level minus 10 dB.

(All these parameters are adjusted to give the time to decay by 60 dB at the relevant slope.)

The first two of the above parameters are defined in ISO 3382 (where they are referred to as T_{20} and T_{30}, respectively) and are commonly used to characterise early reverberation in rooms. EDT explicitly includes the effect of direct sound from the source and is used to characterise the prominence of that sound, similarly to the "clarity" index described below.

In a Sabine space, values of RT_{20} and RT_{30} will both be equal to RT as defined above, and they will be given by the Sabine reverberation time. (EDT will differ as it explicitly includes direct sound.) The actual values of all parameters can be calculated directly from the impulse response. Values for the example spaces shown in Fig. 2.6 are given in Table 2.1.

Clarity and Gain

Two other descriptors that can be used in an attempt to quantify the subjective experience of sound in a space are:

- Clarity, or C_{50}: the ratio of sound energy arriving at the receiver less than 50 ms after the direct sound to energy arriving after 50 ms, expressed in decibels. Sound arriving before 50 ms is found to contribute to a sense of presence or closeness to the source,
- Gain, or G, also known as strength: the absolute sound level of a constant source at the receiver location compared with the sound level of the same source in an anechoic space at 10 m. This is important in spaces such as auditoria where sound levels from unamplified sources need to be sufficiently high at large distances from the source.

Once again, these parameters can be calculated from the impulse response. However, calculation of G requires a scaled response and hence cannot be performed for the unscaled example responses. Values of C_{50} for the example responses are shown in Table 2.1.

Audio Simulations

Impulse responses can be used to create simulations of sound as it would be heard in the space, when the source and receiver are at the relevant locations, by convolving a "dry" recording with the impulse response.

Each of the above impulse responses has been used to simulate speech in the space, using a speech sample from the LibriSpeech dataset produced by OpenSLR. This is described in (Panayotov et al. 2015).

The processed speech samples can be accessed by clicking the links in Table 2.1. The sounds in each space can be compared with the parameters in Table 2.1 and are discussed in the main text, Chap. 8.

Appendix 3: SPL Versus SWL in a Room: Details

SPL and SWL in General

It is often necessary to determine the Sound Pressure Level (SPL) in a space resulting from a source with a given Sound Power Level (SWL), and this Appendix addresses that question. It should first be noted that where sound power propagates away from an omnidirectional source, the general relationship between SPL and SWL can be formulated as

$$\begin{aligned}\text{SPL} = \text{SWL} \\ - 10\ \text{lg}\left(\text{Area over which sound power is spread, in } m^2\right)\end{aligned} \tag{2.14}$$

In the case of spherical spreading from a source, at a distance r metres from the source, the sound power is spread over the surface area of a sphere of radius r metres, and hence,

$$\text{SPL} = \text{SWL} - 10\ \text{lg}\left(4\pi r^2\right) \tag{2.15}$$

For propagation in a duct with cross-sectional area A square metres,

$$\text{SPL} = \text{SWL} - 10\ \text{lg}(A) \tag{2.16}$$

For a source in an enclosed space, these formulae do not apply because the sound power is not propagated away but remains in the enclosure. The relationship between SPL and SWL then depends on properties of the enclosure.

Using Impulse Response

In theory, the difference between SPL and SWL can be calculated directly from the (scaled) impulse response of the space, as described in Appendix 2, if that can be measured or calculated for the required locations of source and receiver.

The impulse response is squared and integrated, ensuring that the integration does not include sections of noise after the impulse has ended, giving a result Δ. The SPL at that location due to a source with sound power level SWL (and the same directional characteristics as the test source) will then be

$$\text{SPL} = \text{SWL} - 10\ \text{lg}\left(4\pi r_0^2\right) + \Delta \tag{2.17}$$

where r_0 is the reference distance for the scaled impulse response.

If using a measured impulse response, the result is very sensitive to the early section of the decay, which in turn is sensitive to the directivity of the source. This should match the directivity of the source whose SPL is required.

In practice it would be unusual to have access to a measured impulse response in a space for which this calculation is required, as the calculation is normally needed at the design stage.

"Sabine" Spaces

If a space is such that reverberant decay of the sound field is linear (in decibels) from very soon after the arrival of direct sound, the sound field can be considered "diffuse", as described in Appendix 2, for the entire decay. In this case, the entire decay can be described by Eq. (2.11) or (2.12). Conditions for this to occur are those listed in Appendix 2, with the further condition that the average absorption coefficient α is small, so that linear decay begins early. Note that if α is small, Eqs. (2.11) and (2.12) give very similar results, and Eq. (2.12) is generally used.

As noted in Appendix 2, spaces where these conditions apply are described as "Sabine" spaces in this book. They include auditoria, plant rooms and other spaces where reverberation is significant. In a Sabine space, the SPL resulting from a given SWL can be calculated by integrating the reverberation curve, giving

$$\begin{aligned} \mathrm{SPL} &= \mathrm{SWL} + 10\,\lg\left(\frac{4}{S\alpha}\right) \\ &= \mathrm{SWL} + 10\,\lg(\mathrm{RT}) - 10\lg(V) + 14 \end{aligned} \tag{2.18}$$

where S is the surface area of the space in m^2, RT is the reverberation time in seconds and V is the volume in m^3, and we have used Eq. (2.12).

This represents the SPL resulting from the reverberant section of the decay. The contribution of direct sound from the source can be added, taking account of any directivity, giving

$$\mathrm{SPL} = \mathrm{SWL} + 10\,\lg\left(\frac{Q}{4\pi r^2} + \frac{4}{S\alpha}\right) \tag{2.19}$$

where Q is the directivity of the source.

It is worth noting that Eq. (2.18), representing the reverberant SPL, does not require that the source be a point source. In particular, it may be a wall, floor or ceiling, and this will be used in Chap. 3 of the main text.

"Schultz" Spaces

If the space of interest does not meet the requirements of a Sabine space, the only generally applicable way to estimate SPL from SWL is to calculate an impulse response using a room modelling programme and use the procedure in Appendix B. However, in most cases, this would be very cumbersome.

Fortunately, many "normal" spaces such as domestic spaces, offices, and cafes that do not meet the requirements of a Sabine space still exhibit regularities in the relationship between SPL and SWL. A paper by Schultz (1985) postulates a general relationship valid for these spaces, namely

$$\mathrm{SPL} = \mathrm{SWL} - 3\,\lg(f) - 5\,\lg(V) - 10\,\lg(r) + 12 \tag{2.20}$$

where f is an octave-band centre frequency in Hz.

This equation is based on no theory at all—it simply summarises the results of multiple measurements of SPL from a standard source in "normal" spaces. Note that it includes an explicit dependence on frequency—in other words, it assumes a "typical" frequency dependence of the absorption in the space. Note also that it includes a dependence on r throughout the space—there is no "critical distance" beyond which the SPL remains constant with distance.

Spaces where this equation is valid are sometimes called "Schultz" spaces. The equation is often used when calculating SPL values from mechanical services in furnished offices, apartments and similar spaces.

Note that Eq. (2.20) does assume that the source is a point source. However if the equation is considered accurate, the SPL from an extended source can be found by integrating the resulting SPL over the source area. This gives rise to a correction, Δ_s, for the size of the source. For a circular source of radius ρ, the correction is found to be

$$\Delta_s = 10\lg\left\{\frac{2}{\overline{\rho}^2}\left[\sqrt{1+\overline{\rho}^2} - 1\right]\right\} \tag{2.21}$$

in which $\overline{\rho} = \rho/r$. This correction is plotted in Fig. 2.10.

In the case of a rectangular source, the integral for the correction does not have an analytical solution. However, the correction can be estimated with an accuracy of better than 1dB by using Eq. (2.21) but replacing

$$\overline{\rho}^2 = \frac{L_1\sqrt{L_1 L_2}}{\pi r^2} \tag{2.22}$$

where L_1 is the length of the longer side and L_2 the length of the shorter side.

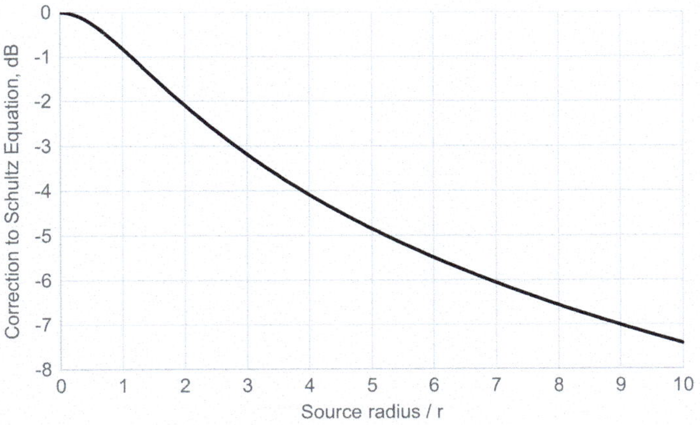

Fig. 2.10 Correction to be added to Eq. (2.20) in the case of a circular source of finite radius

Limitations

Equations (2.19) and (2.20) both have limitations as to the spaces and locations where they are valid. The major limitations are as follows.

First, in small rooms at low frequencies, room modes contribute significantly to the SPL at certain frequencies. These are described in Chap. 8 of the main text. Their effect cannot be accounted for even when using standard room modelling computer programmes employing "ray tracing". A full description requires a more complicated "finite element" model, although techniques described in (Hopkins 2007) can identify the relevant frequencies.

The lowest frequency at which Eqs. (2.19) and (2.20) are valid can be estimated by the "Schroeder cut-off frequency"

$$f_{\text{cut - off}} = \frac{840}{\sqrt{S\alpha}}$$

For small spaces of about 60 m^3, this limit is likely to be at least 200 Hz. For spaces of about 300 m^3, it is unlikely to be above 125 Hz.

Second, at receiver locations close to a reflecting surface, the direct and reflected sound may add in phase, giving an increase of up to 6 dB in the direct sound component of SPL. To avoid an increase of more than 1 dB in SPL, using conservative assumptions, the receiver should be located at a distance greater than 0.15λ from a reflecting surface, where λ is the wavelength of the lowest frequency of interest.

References

Acoustical Society of America (1997) ASA/ANSI Standard S3.5-1997 methods for calculation of the speech intelligibility index. Acoustical Society of America

Acoustical Society of America (2023) ASA/ANSI standard S1.42-2023: design response of weighting networks for acostiocal measurements. Acoustical Society of America

Hopkins C (2007) Sound insulation. Elsevier

International Standards Organisation (2008a) ISO 3382-2: acoustics-measurement of room acoustic parameters—part2: reverberation time in ordinary rooms. International Standards Organisation

International Standards Organisation (2008b) ISO 3382-1 acoustics-measurement of room acoustic parameters—part 1: performance rooms. International Standards Organisation

International Standards Organisation (2019) ISO 3740:2019 acoustics—determination of sound power levels of noise sources—guidelines for the use of basic standards. International Standards Organisation

Merimaa J, Peltonen T, Lokki T (2005) Concert hall impulse responses—Pori, Finland: Reference. Retrieved from http://legacy.spa.aalto.fi/projects/poririrs/docs/poriref.pdf

Moore B (2013) An introduction to the psychology of hearing, 6th edn. BRILL

Panayotov V, Chen G, Povey D, Kudanpur S (2015) Librispeech: an ASR corpus based on public domain audio books. IEEE International conference on acoustics, speech, and signal processing

Pierce A (2019) Acoustics: an introduction to its physical principles and applications. Springer Cham

Randall R (2005) An introduction to acoustics. Dover Publications

Schultz T (1985) Relationship between sound power level and sound pressure level in dwellings and offices. ASHRE Trans 91(1A):124–153

Standards Australia; Standards New Zealand (2002) AS/NZS 2460:2002 acoustics—measurement of reverberation times in rooms

Szoke I, Skacel M, Mosner L, Paliesek J, Cernocky J (2019) Building and evaluation of a real room impulse response dataset. IEEE J Sel Top Sign Process 13(4):863–876

Chapter 3
Sound Power Level Through a Partition

Abstract This Chapter describes how sound travels through a partition. It introduces the sound reduction index R and shows how the sound power entering a space on the receiver side is related to R as well as to properties of the sound field on the source side. Associated spreadsheets provide ways to automate the calculations.

3.1 Introduction

In this book, a *partition* is taken to be any material that divides a space from another area where separate activities are conducted. The adjacent area could be another internal space or the external environment.

When sound enters a space through a partition, the sound power entering depends on:

- the sound level in the adjacent area,
- properties of the partition and
- properties of the sound field in the adjacent area.

This chapter describes how to predict the sound power that will enter the space.

3.2 The Sound Reduction Index *R*

The *transmission coefficient* τ of a partition can be defined as the ratio of the sound power radiated by the partition on the receiving side (W_2) to the power incident on the source side (W_1). (Equivalently it can be defined as a ratio of sound intensities.) The *sound reduction index* R is defined as

Supplementary Information The online version contains supplementary material available at https://doi.org/10.1007/978-981-97-9380-8_3.

R. Bullen, *The Architecture of Sound*, Springer-AAS Acoustics Series,
https://doi.org/10.1007/978-981-97-9380-8_3

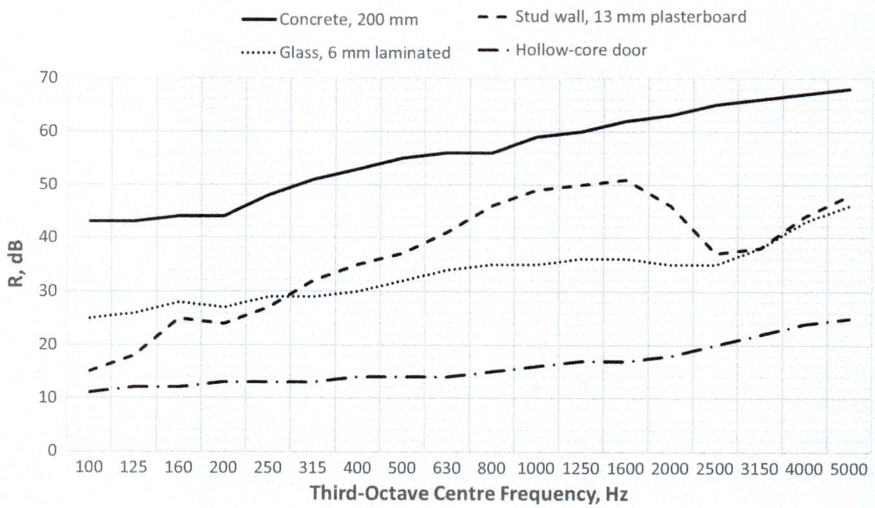

Fig. 3.1 Values of *R* for some common building elements

$$R = -10\lg(\tau) = 10\lg(W_1) - 10\lg(W_2) = \mathrm{SWL}_1 - \mathrm{SWL}_2 \tag{3.1}$$

("Sound reduction index", or *R*, is the term used for this unit in international standards, and that terminology will be used in this book. However, the term "transmission loss", often abbreviated as TL, is also used and has the same meaning.)

R always depends on the frequency of the incoming sound. Figure 3.1 shows examples of the sound reduction index for some common building elements. (As explained below these are technically the "random-incidence sound reduction index", R_{rand}). As seen in Fig. 3.1, *R* is generally higher—that is, the sound power propagated through the material is lower—at higher frequencies.

3.3 Sound Power Transferred Through a Partition

Referring to Fig. 3.2, if an incoming sound is a plane wave in a free field, with sound pressure level $\mathrm{SPL}_{\mathrm{in}}$, then the sound power incident on the partition is

$$\mathrm{SWL}_1 = \mathrm{SPL}_{\mathrm{in}} + 10\lg(S\cos\theta) \tag{3.2}$$

where

- S is the area of the partition, measured in m^2, and
- θ is the angle between the incoming plane wave and the normal to the partition.

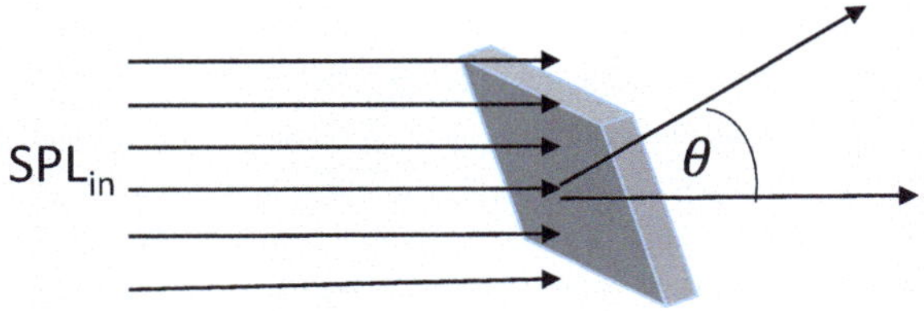

Fig. 3.2 A plane wave incident on a partition, at an angle θ

Note that SPL_{in} is *not* the sound pressure level that would be measured by a microphone placed in front of the partition—it is the SPL that would be measured at that point if the partition were not there. The partition will produce reflected sound, and close to the partition this will interfere with the direct sound, creating increases in measured SPL of up to 6dB at some frequencies and large decreases at other frequencies. This reflected sound is directed away from the partition and does not result in increased sound propagation through it.

The partition then represents a surface source of sound power for the space on the receiving side, with sound power level

$$SWL_{surf} = SPL_{in} + 10\lg(S\cos\theta) - R(\theta) \tag{3.3}$$

As indicated in this equation, in general R depends on the angle of incidence θ.

The sound power transmitted clearly depends on the range of angles at which sound strikes the partition. Five cases cover most situations encountered in practice.

1. **A plane wave striking the panel at close to normal incidence**—for example a single vehicle at its closest approach to an external wall. In this case $\theta = 0$ and Eq. (4.3) becomes:

$$SWL_{surf} = SPL_{in} + 10\lg(S) - R(0) \tag{3.4}$$

   $R(0)$ is known as the *normal-incidence sound reduction index.*
2. **Plane waves arriving from a range of angles**—for example external sound from a stream of vehicles. In this case, the following approximation is often used:

$$SWL_{surf} = SPL_{in} - 3 + 10\lg(S) - R_{field} \tag{3.5}$$

   where R_{field} is the weighted average value of $R(\theta)$ for the relevant angles and is known as the *field-incidence sound reduction index.*
3. **If the source is in a Sabine room** as described in Chap. 2, that is, one where the sound forms a diffuse field, and the source is not close to the partition, the sound power incident on the partition will be equal at all angles. In this case, Eq. (3.3) is replaced with

$$SWL_{surf} = SPL_{rev} - 6 + 10\lg(S) - R_{rand} \tag{3.6}$$

where SPL_{rev} is the SPL in the section of the source room where a diffuse field exists (not close to the partition) and R_{rand} is the *random-incidence sound reduction index*. $SPL_{rev} - 6$ represents the proportion of the sound energy in the reverberant field that is directed into the partition.

For a single source in a Sabine room, SPL_{rev} is given by Eq. (2.1).

4. **If the source is in a Schultz room** (see Chap. 2), the SWL propagated through the partition is estimated as

$$SWL_{surf} = SPL_{av} - 3 + 10\lg(S) - R_{field} \tag{3.7}$$

where SPL_{av} is the (logarithmic) average SPL incident on the partition. For a single source in a Schultz room, SPL_{av} is given by Eq. (2.5).

5. **If a single point source is very close to the partition** (compared with the dimensions of the partition) either externally or in an adjacent space, the above procedures break down. In this case, a reasonable estimate of the sound level in the receiving space can be found by the following process.

Replace the source with a "virtual" point source in the receiving space, on the other side of the partition. If the sound power level of the real source is SWL_{real}, the sound power level of the virtual source is

$$SWL_{virtual} = SWL_{real} + 10rmlg(Q(\theta)) - 3 - R_{rand} \tag{3.8}$$

where θ represents the direction from the source to the partition. The directivity of the *virtual* source is 2. That is, for the virtual source $10\lg(Q) = 3$ at all points in the receiving space.

3.4 Dependence of *R* on Angle of Incidence

Figure 3.3, derived from Brutel-Vuilmet et al. (2006), shows measured values of $R(\theta)$ for various angles of incidence, as well as R_{rand}, for a panel of 10 mm glass. These values will also depend on the size of the panel and the manner of fixing, although R_{rand} is less susceptible to those influences than individual $R(\theta)$ values.

The variation in R with angle of incidence is significant at mid to high frequencies. R_{rand} tends to be lower than $R(\theta)$ except at large values of θ—that is, at near-grazing incidence.

In practice, laboratory measurements of $R(\theta)$ are very difficult to perform, while measurement of R_{rand} is considerably simpler. Even if values of $R(\theta)$ were known, as in Fig. 3.3, their use in predictions would require detailed knowledge of the distribution of incident sound power by angle, which is generally not available.

For these reasons, measurement procedures described in standards including AS 1191–2002 Australian Standards Association (2002) are for the measurement of R_{rand}, and almost all values of R that are quoted in standard data sheets represent the random-incidence sound reduction index. In the absence of information on the

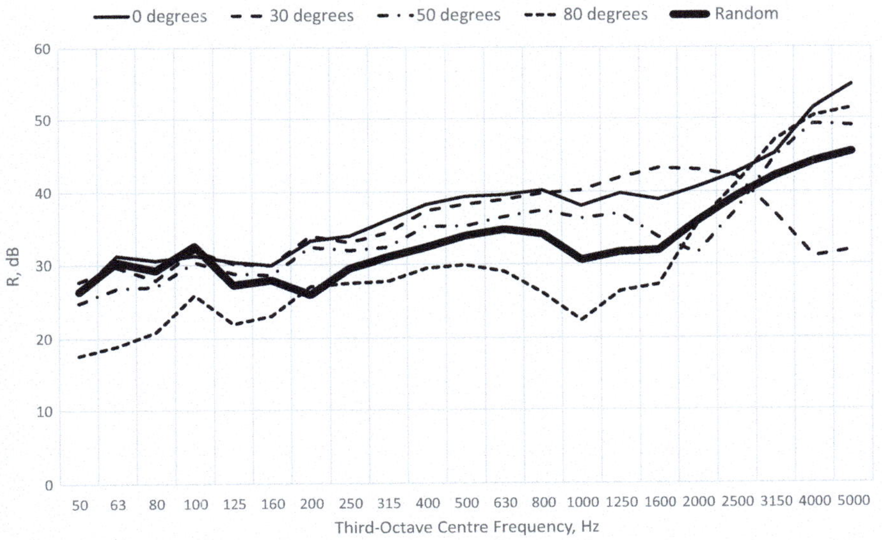

Fig. 3.3 Measured values of $R(\theta)$ at incidence angles of 0°, 30°, 50° and 80°, as well as R_{rand} for single 10mm glazing—adapted from Brutel-Vuilmet et al. (2006)

variation of $R(\theta)$ with angle and/or the angular distribution of noise from the source, Eqs. (4.4)–(4.8) are used with the assumption $R(0) = R_{\text{field}} = R_{\text{rand}}$. This almost always gives a conservatively high estimate of the sound power transmitted into the receiving space.

In general literature, when the "sound reduction index" R is referred to, this will invariably mean the random-incidence index R_{rand}, and this convention will be followed in subsequent sections of this book. Equations (3.4)–(3.8) can be used with R substituted for $R(0)$, R_{rand} and R_{field}.

Exercise 3.1

(a) Supplementary file01 gives a spectrum of noise from an aircraft at the location of a proposed residence, and R values for the roof/ceiling system. Calculate the A-weighted SWL entering a room of the residence through a 4 m × 5 m ceiling.

(b) A person is speaking in a raised voice in a Sabine room with volume of 150 m^3 and reverberation time 0.8s at all frequencies. There is a 4 m × 2.8 m stud wall with the R values given in this chapter spreadsheet between this and another space. What is the A-weighted SWL transferred through the wall?

(c) Repeat (b) for a Schultz room, volume of 150 m^3, with the talker 3 m from the wall and facing the wall.

(d) If a person speaks in a raised voice very close to the wall, calculate the A-weighted SWL of a virtual source in the receiving room, if the voice is directed straight at the wall.
(e) What would be the A-weighted SWL if they were speaking directly away from the wall?

3.5 SPL in the Receiving Space

We can now calculate the SPL in a receiving space using the SWL values calculated through Eqs. (3.4) to (3.8) and the conversions to SPL in Chap. 2.

The simplest case is transmission from a plane wave source to a Sabine room. The SWL transferred is given by Eq. (3.4), remembering that SPL_{in} is the outside SPL *in the absence of the partition.* The SPL in the room is then given by Eq. (2.7) (because the source of SWL in the room is a surface). The combined result is:

$$SPL_{receiver} = SPL_{in} + 10\lg(S) - R + 10\lg(T) - 10\lg(V) + 14 \tag{3.9}$$

where T and V are respectively the reverberation time and volume of the receiving room.

Similar combined equations can be written for other combinations of source and receiver spaces. For example, for transmission from a Sabine room to another Sabine room, Eqs. (3.6) and (3.7) give:

$$SPL_{receiver} = SPL_{rev} + 10\lg(S) - R + 10\lg(T) - 10\lg(V) + 8 \tag{3.10}$$

This configuration is used for measuring the R values of materials by rearranging Eq. (3.10) to give R in terms of the other variables.

In setting out calculations, it is often helpful to explicitly calculate the SWL transmitted into the receiver space from all paths, using Eqs. (3.4)–(3.7), and then use these values to calculate the SPL in the space using Eqs. (2.7)–(2.11) (or Chap. 2 spreadsheet). This makes the calculation procedure clear and sometimes avoids making erroneous assumptions.

Of course, if sound enters a space via two or more identifiable paths, then the results need to be added on a logarithmic basis to give the total SPL in the receiving space.

Another example is intrusion of external sound through an open window. If we assume a receiver room of 70 m^3 with external sound incident on a partially open window (say open area 1 m × 0.6 m, with $R = 0$) then for a Sabine space with $T = 0.7$, Eqs. (3.5) and (2.7) give:

$$
\begin{aligned}
SPL_{Sabine} &= \left[SPL_{in} - 3 + 10\lg(0.6) - 0\right] + 10\lg(0.7) \\
&\quad - 10\lg(70) + 14 \\
&= SPL_{in} - 11
\end{aligned}
$$

For a Schultz space, with the receiver at 1 m from the window and a sound centred at 250 Hz, Eqs. (3.5) and (2.11) give:

$$SPL_{Schltz} = SPL_{in} - 10$$

This leads to the common assumption that in a residential building, a partially open window provides a reduction of about 10 dB in the external sound level. (Note once again that this refers to a reduction compared with the external sound level in the absence of any increase due to reflection from the building façade.) However, this calculation includes several assumptions including the size of the opening, the size of the room and sound incidence on the window being at a range of angles so that Eq. (2.5) is used rather than Eq. (2.4). These should all be verified before accepting this simple formula.

Exercise 3.2

(a) Return to Exercise 3.1, part (c) (person talking in a Schultz room). A listener is in an adjacent room, 2 m from the wall. This is a Schultz room, with volume of 70 m^3. What is the A-weighted SPL at the listener position?

(b) A person is speaking in a raised voice close to the stud wall, facing towards the wall, as in Exercise 3.1, part (d). The listener in the Schultz room is 3 m from where the speaker is standing. What SPL will they hear through the wall, in dBA?

References

Australian Standards Association (2002) Acoustics—method for laboratory measurement of airborne sound insulation of building elements. Australian Standards Association

Brutel-Vuilmet C, Guigou-Carter C, Villot M, Jean P (2006) Measurement of the sound reduction index as a function of the incidence angle by two different methods. Building Acoustics 13:33–48

Chapter 4
Sound Power Level Through a Duct

Abstract This chapter describes propagation of sound through a duct, with particular reference to air-conditioning ducts which represent the most common use case for these procedures. It covers generation of sound from fans and from other items in the duct; attenuation by duct lining, silencers and other methods and calculation of the sound power entering a receiving space. Associated spreadsheets provide ways to automate the calculations.

4.1 Introduction

Ducts convey various services between spaces throughout a building and have the potential to transmit sound between those spaces. The most prominent example of this is sound transmission through air-conditioning ducts, and this will be the focus of this chapter. Transmission through other types of ductwork can be predicted and controlled in much the same way as sound in air-conditioning ducts.

4.2 Noise from a Large Air-Conditioning System

In an air-conditioning system for a large building, temperature-controlled air is sent to all relevant spaces within the building by a fan, located in a plant room somewhere in the building. Air travels to spaces in the building through ducts (the "supply air" system) and enters each space through outlets or "diffusers", generally in the ceiling. Air returns to the fan through other ducts (the "return air" system) and:

- mixes with fresh air—often in a box known as a "plenum",
- has its temperature controlled by a chiller or heater and,

Supplementary Information The online version contains supplementary material available at https://doi.org/10.1007/978-981-97-9380-8_4.

R. Bullen, *The Architecture of Sound*, Springer-AAS Acoustics Series,
https://doi.org/10.1007/978-981-97-9380-8_4

- is sent again through the supply-air system.

The "air handling unit" consists of the fan and associated filters and coils for temperature control. Figure 4.1 shows simplified diagrams of these two systems.

The major source of noise in this process is the fan, and this noise can be transmitted to internal spaces in three ways:

- along the ductwork together with the air. Note that although in the return air system air is moving from the space to the fan, *noise* can still move from the fan to the space,
- through a wall, in the case of spaces directly adjacent to the plant room,
- by vibration generated by the fan and passed through the building structure.

Noise can also be generated by the air flow itself and by other components of the system located along the ducts.

Noise passing through a plant room wall is a special case of room-to-room sound transmission as discussed in Chap. 3.

Noise generated by vibration is discussed in Chap. 7.

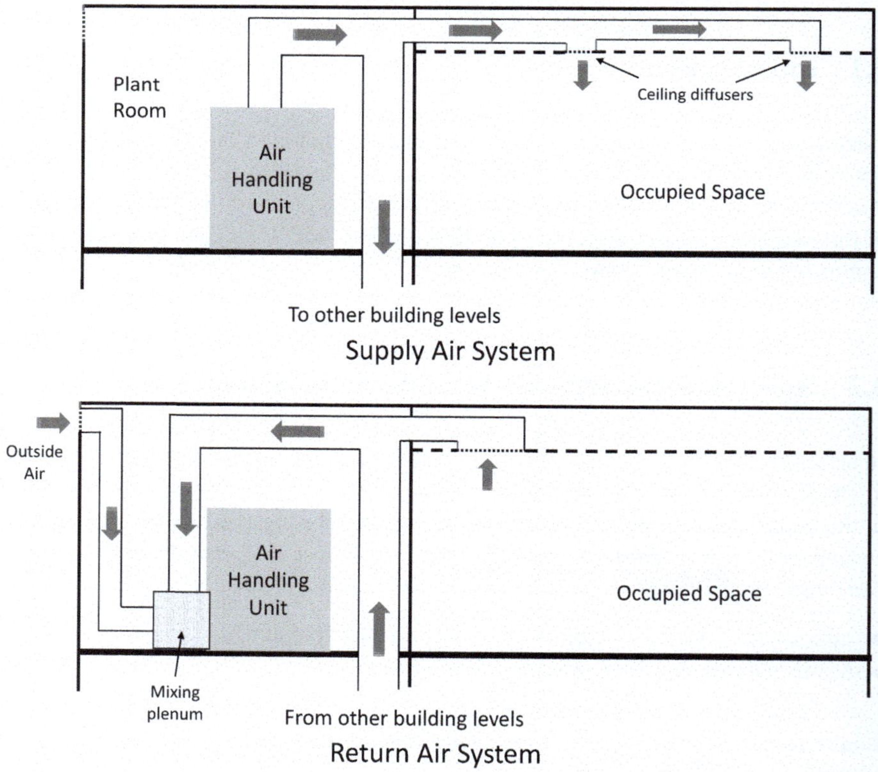

Fig. 4.1 Simplified schematic diagram of the supply air and return air systems for a large building. Arrows show the direction of air flow

This chapter will describe the calculation of sound levels due to sound passing down the ducts, which in practice is usually the major source of sound from these systems.

The best reference book describing noise from building services, including air-conditioning noise, is the ASHRAE handbook produced by the American Society of Heating, Refrigeration and Air-conditioning Engineers, Chapter 48: Sound and Vibration Control (ASHRAE 2019). This should be consulted for detailed information on noise from building services. Another useful reference is the CIBSE Guide, published by the Chartered Institution of Building Services Engineers, London (Chartererd Institution of Buildng Services Engineers 2005).

The information below summarises procedures for calculating "down-duct" noise from building air-conditioning systems.

4.3 Sound Power Level from the Fan

We begin with the sound power level (SWL) injected into a duct by the fan. This depends on:

- the size and type of the fan,
- the pressure against which the fan is pushing, which depends on the size of the building and the layout of the ductwork.

The two major types of fan are shown in Fig. 4.2. Axial fans are used for low-pressure applications such as exhaust fans. Fans supplying air for a large building are invariably centrifugal.

The mechanical services consultant who specifies the fan should be able to provide the sound power levels injected by their fan into the supply and return air systems,

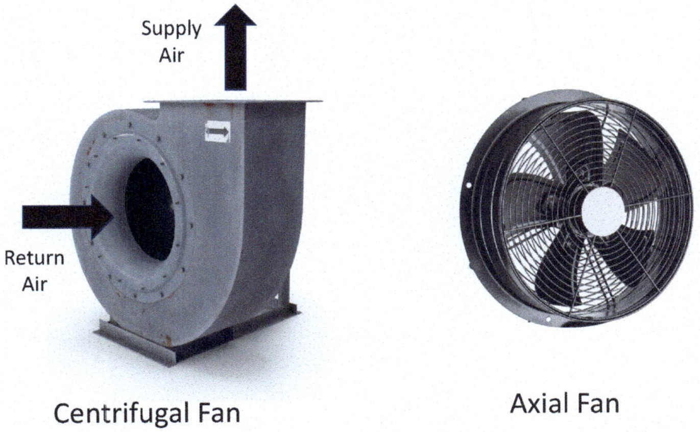

Fig. 4.2 Major types of fan

as well as the sound power level radiated by the fan into the plant room. These will usually be in octave bands and are all you need to begin your calculation.

4.4 Attenuation by the Duct

Sound power is attenuated to a small extent by unlined ductwork, primarily at low frequencies. Attenuation values in dB per metre of duct are provided in the Supplementary file1 spreadsheet.

However, much higher attenuation is achieved when the duct is lined with absorbing material. This can be fibreglass or similar material and is typically 25 mm or 50 mm thick, usually with a facing of perforated foil.

In this case, attenuation is measured by "insertion loss"—i.e. the difference between the SWL with and without the absorptive lining. In theory the total attenuation would equal the "insertion loss" plus the attenuation from unlined duct. However, the attenuation of unlined straight duct is so small that it is generally ignored in calculations.

Appendix 1 gives details of the recommended procedure to calculate the insertion loss from lined duct, which comes from (ASHRAE 2019). The procedure is complicated, but the Supplementary file1 spreadsheet includes a calculation sheet which performs the calculations for you.

Figure 4.3 shows an example of the attenuation calculated for a 610 × 610 mm rectangular duct 3 m long and lined with 25 mm or 50 mm absorbing material.

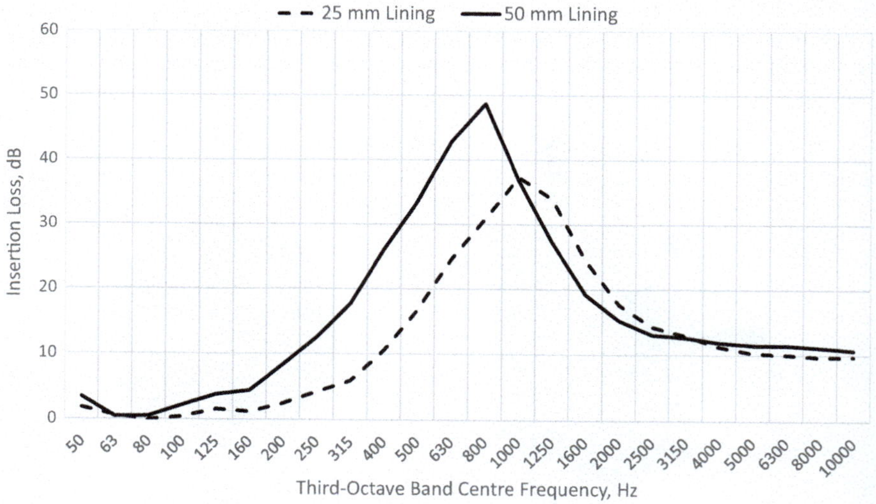

Fig. 4.3 Typical insertion loss for lined rectangular ducts—610 × 610 mm duct, 3 m length (data from ASHRAE (2019))

Exercise 4.1
The SWL values, in dB, injected by a fan into a duct are given as follows:

Octave-band centre frequency, Hz						
63	125	250	500	1 K	2 K	4 K
84	86	80	77	73	70	65

(a) Calculate the A-weighted value of this SWL, as described in Chap. 2.
(b) Using attenuation values from the Supplementary file1 spreadsheet, calculate the A-weighted SWL after passing through 3 m of unlined 305 × 610 mm duct
(c) Calculate the A-weighted SWL after 3 m of the same duct lined with 25 mm thick absorbing material.

4.5 Duct Silencers

If greater attenuation is required, a silencer can be placed in the duct. These are devices that effectively divide the duct into several acoustically lined passages. Figure 4.4 shows a typical silencer.

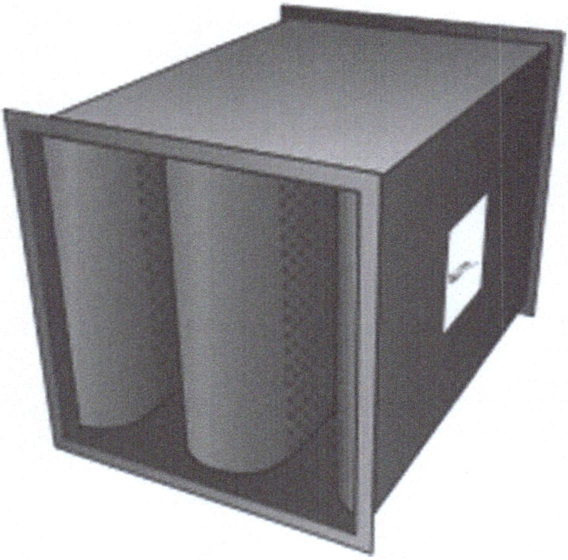

Fig. 4.4 Typical duct silencer Image courtesy of IAC Acoustics

Insertion loss values for some typical silencer configurations, taken from (IAC Acoustics n.d.), are included in the Supplementary file1 spreadsheet. These values give some idea of achievable insertion loss values, but in specifying a silencer careful reference should always be made to the manufacturer's supplied data.

Exercise 4.2
For the fan in the previous question, calculate the A-weighted SWL at the end of the duct if the duct lining is replaced by a 1200 mm silencer, using silencer transmission loss data from the in the Supplementary file1 spreadsheet.

4.6 Duct Branching

When ducts branch into two or more sub-branches, sound power is distributed in proportion to the internal area of the duct in each branch. For example, if branch 1 has area A_1 and branch 2 has area A_2, the SWL into branch 1 is

$$SWL_1 = SWL_{total} + 10\lg\left(\frac{A_1}{A_1+A_2}\right) \tag{4.1}$$

In particular, if $A_1 = A_2$, then the sound power in each branch is $SWL_{total} - 3$. This is true at all frequencies.

Exercise 4.3
The total SWL in a duct is 80 dB. The duct splits into two, with the area of the first new duct being twice the area of the second. What is the SWL in the second of the new ducts?

4.7 Attenuation at Duct Outlets

When a duct opens into a space, some sound power is reflected back into the duct and does not reach the space. This process is descried in detail in (Bradflo 2024). Unlike most losses, this occurs largely at low frequencies. Figure 4.5 shows typical losses from circular duct openings. Data for other duct sizes is included in the Supplementary file1 spreadsheet. For other shaped openings, the losses can be approximated by those from a circular opening with the same area.

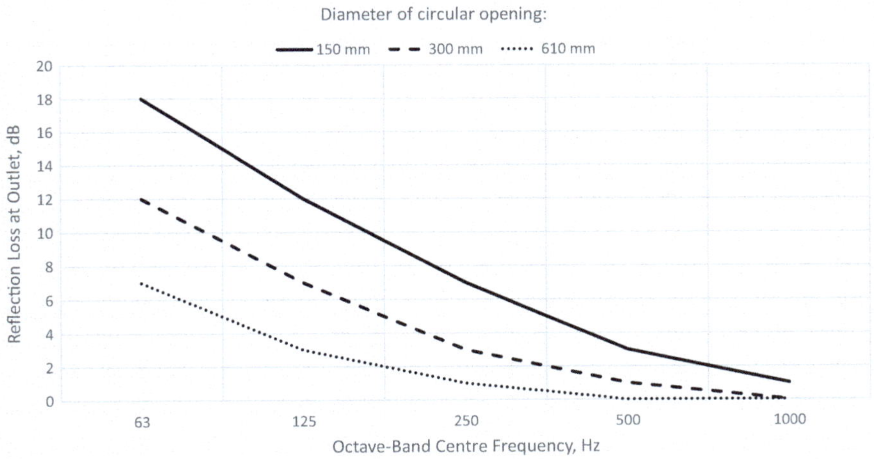

Fig. 4.5 Typical end reflection loss values for circular ducts of various diameters (data from ASHRAE (2019))

4.8 Attenuation by a Plenum

A plenum is a large chamber used to smooth airflow. They are commonly found on the return air side of an air-conditioning system between the fan and the ducted system. Plenums are typically lined with absorbing material and can provide useful sound attenuation. Figure 4.6 shows a schematic view of a plenum.

The attenuation provided by a plenum is described by the transmission loss—the difference between the SWL in the inlet duct and the SWL in the outlet duct. ASHRAE (2019) gives a method of estimating the transmission loss which is quite complex and is implemented in the relevant sheet in the Supplementary file1 spreadsheet. The method requires a number of parameters, including:

- volume of the plenum,
- dimensions of the inlet and outlet ducts,
- thickness of absorptive lining,

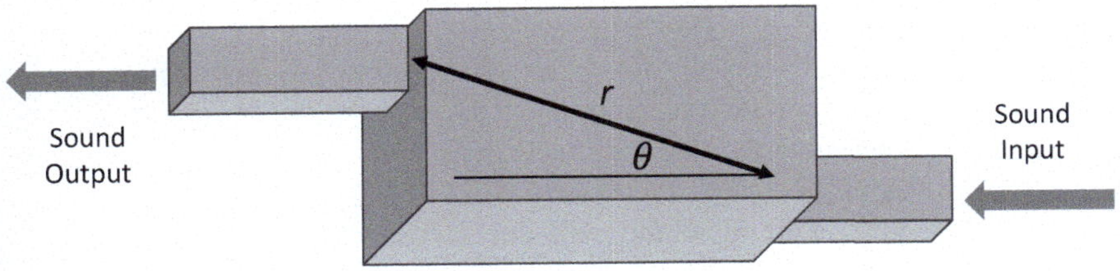

Fig. 4.6 Schematic view of an air-conditioning plenum

- surface area of the lining—that is, the surface area of the *inside* of the lining, which does not include the areas of the inlet and outlet ducts,
- inlet duct position (close to a wall of the plenum or not),
- distance between centres of inlet and outlet ducts (r in Fig. 4.6),
- angle between inlet and outlet ducts (θ in Fig. 4.6).

Note that here "inlet" and "outlet" refer to the inlet and outlet of *sound*. If the plenum is in the return air system, *air* will be flowing in the opposite direction to the sound.

Exercise 4.4

Use the Supplementary file1 spreadsheet to calculate the transmission loss of the plenum shown in Fig. 4.6, in third-octave bands, given the following information.

Plenum size: 1500 × 800 × 800 mm.

Inlet duct: 400 × 500 mm, not close to a wall.

Outlet duct: 500 × 600 mm.

Absorptive lining: 50 mm.

Distance between centres of inlet and outlet ducts: 1580 mm.

Angle between centres of inlet and outlet ducts: 30°.

4.9 Flexible Duct (also Known as "Flexiduct" or "Flexduct")

Lined flexible ducts are commonly used for the final section of a supply-air system, between a metal duct and a ceiling outlet. Insertion loss values for typical sizes and lengths of lined flexible duct are provided in the Supplementary file1 spreadsheet, and these values can be subtracted from the SWL reaching the outlet exactly as for metal ducts.

However here there is a complication because unlike metal ducts, sound can easily break out of flexible ducts and travel directly to the space below through the ceiling. The sound power reaching the space in this way depends on the sound reduction of the ceiling. The two sound paths are illustrated in Fig. 4.7.

In path 1, the SWL at the end of the metal duct is reduced by the attenuation of the flexible duct and reflection loss from the outlet and then enters the occupied space at the outlet.

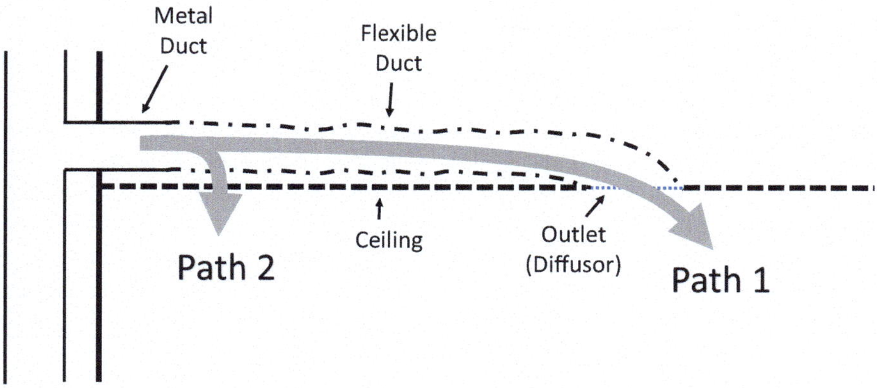

Fig. 4.7 Sound paths from flexible duct into an occupied space

In path 2, the SWL leaves the flexible duct and is reduced by the sound reduction *R* of the ceiling before entering the occupied space beneath the end of the metal duct. (This is equivalent to Eq. (3.8) for room-to-room transmission where we assume all power is directed through the ceiling, so $10\lg(Q(\theta)) = 3$).

Sound reduction values for various materials are discussed in detail in Chap. 6. At this stage, the values of *R* for various ceiling systems can be assumed to be given by the manufacturer, in octave or third-octave bands.

How much sound power travels through the duct and how much breaks-out depends on details of the system and is difficult to predict. Which path will represent the "worst case" in terms sound transmission into the space is also not obvious and depends in particular on the ceiling used.

A common practice in this situation is to perform two calculations, one assuming all SWL travels via path 1 and one assuming it all travels via path 2. The assumed sound level in the space will be the higher of the two values.

Exercise 4.5

Sound power with levels as specified in the table below enters a flexible duct, 2.7 m long and 200 mm in diameter, which discharges air through a 200 mm diameter outlet. Sound reduction values for the ceiling are also given in the table. Which of paths 1 and 2 will produce the highest A-weighted sound power entering the space below?

	Octave band centre frequency, Hz						
	63	125	250	500	1 K	2 K	4 K
SWL entering flexible duct, dB	78	78	74	69	60	55	45
Sound reduction *R* of the ceiling, dB	22	28	27	28	34	48	53

4.10 Break-Out from Metal Ducts

More rarely, you may need to consider sound power breaking out from metal ducts. This is typically an issue where a long stretch of duct passes over a ceiling. As in path 2 in the previous section, the SWL reaching the space depends on the R value of the ceiling, but in this case, it also depends on a parameter related to the duct, known as TL_{out}, which represents how much sound power leaves the duct. Values of TL_{out} for typical duct sizes are given in the Supplementary file1 spreadsheet.

Assuming the duct is close to the ceiling, the SWL reaching the occupied space is approximately

$$\mathrm{SWL}_{\mathrm{breakout}} = \mathrm{SWL}_{\mathrm{in\text{-}duct}} + 10\,\mathrm{lg}\left(\tfrac{S}{A}\right) - \mathrm{TL}_{\mathrm{out}} - R \tag{4.2}$$

in which

S = area of duct adjacent to the ceiling. That is, $S = XL$, where X is the width of duct adjacent to the ceiling and L is the length of exposed duct. (In the rare case of circular metal ducts X can be taken as the duct diameter.)
A = cross-sectional area of the duct.
(ASHRAE 2019 gives a more accurate equation).

Note that on entering the space this sound power source represents a surface, whereas the other sources described in this chapter represent points.

Exercise 4.6
Return to the result of Exercise 4.2 (fan noise through a 305 × 610 mm duct with silencer). After the silencer, the duct passes for 6 m over a ceiling with the R values shown in Exercise 4.5. The long side of the duct is next to the ceiling. What is the A-weighted sound power propagated into the room below due to duct breakout?

4.11 Variable Air Volume (VAV) Boxes

These items are often placed above the ceiling in the supply-air system to provide local control over air volumes independent of the main system (Fig. 4.8). They may reduce noise from the main fan slightly as it propagates through the box. However, many VAV boxes incorporate a fan which adds both in-duct and case-radiated sound power, both of which will be specified by the manufacturer.

The in-duct SWL from the VAV box needs to be logarithmically added to the SWL at the input to give the SWL in the subsequent ductwork.

Fig. 4.8 Typical layout of fan-assisted VAV box

The case-radiated sound power can be considered as for breakout noise—the SWL entering the space below is the case-radiated SWL minus the sound reduction of the ceiling.

4.12 Regenerated Noise

Components that involve an intrusion into a duct, or a sudden change in configuration, such as silencers, diffusers and VAV boxes, generate additional SWL through the creation of turbulence in the air stream. The SWL generated depends strongly on the air velocity. The spectrum generally has more high-frequency content than other noise sources and may therefore be useful in creating masking sound if that is desirable. However, for the same reason it can cause loss of speech intelligibility in situations where intelligibility is critical.

The components for which regenerated noise is most significant are silencers and diffusers. Manufacturers of these components will generally supply the SWL values due to regenerated noise, which will always depend on the velocity of air in the duct and often on other parameters. Figure 4.9 shows how typical A-weighted SWLs from regenerated noise vary with flow velocity. Some typical spectra for silencers and diffusers are provided in the Supplementary file1 spreadsheet. Flow velocity, of course, depends strongly on the duct area, so where regenerated noise is significant, larger ducts are generally required.

Regenerated noise levels should be logarithmically added to the in-duct noise at the point where they are created. In the case of diffusors, this is immediately before entry into the occupied space.

Exercise 4.7

The diagram below shows a section of ductwork. The SWL entering from the left and the SWL added by the VAV box are given in the table below. The area of the duct going to the diffuser is 30% of the area of the duct that continues. The diffuser is 200 mm diameter and the air velocity through it is 5 m/s.

What is the A-weighted SWL from the diffuser, including both down-duct and regenerated noise?[1] Which of these sources of SWL is more significant at low frequencies, and which at high frequencies?

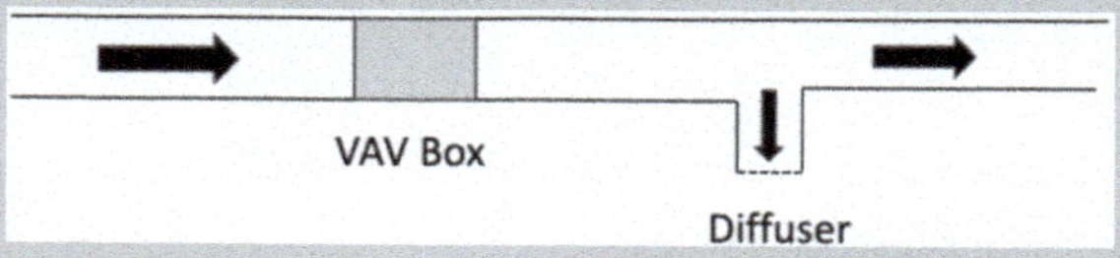

	Octave band centre frequency, Hz						
	63	125	250	500	1 K	2 K	4 K
SWL entering this section of duct	78	78	74	69	60	55	45
Down-duct SWL from VAV box	78	76	71	63	57	52	49

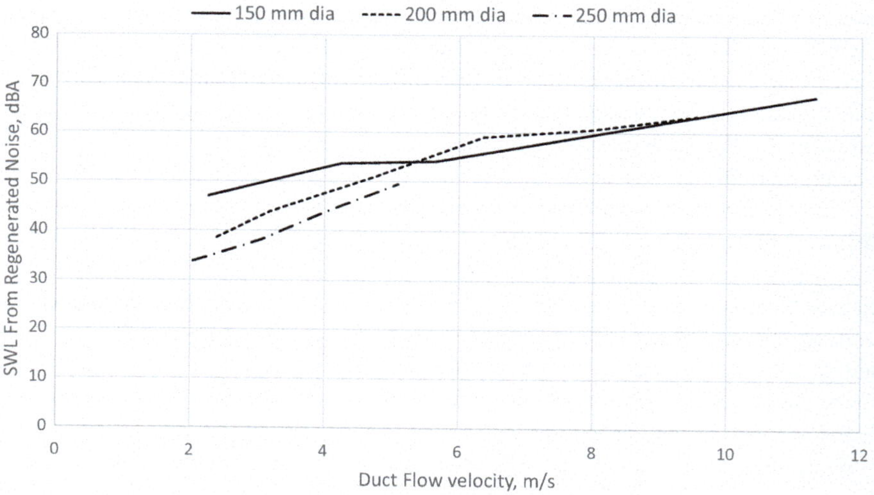

Fig. 4.9 A-weighted regenerated noise from typical ceiling diffusors as a function of air velocity in the duct (data from Bradflo (2024))

[1] Use regenerated noise levels from the Supplementary file1 spreadsheet, and don't forget reflection losses at the outlet.

4.13 A Final Complication—Pressure Drop

The volume of air that any fan can move depends on the "pressure drop" of the system it is pushing the air into. This is a measure of the resistance to air flow created by ducts, various devices in the duct and the rooms themselves. Increasing the pressure drop results in lower air flow, until at a certain point the fan begins to create significant noise and eventually stalls. This is shown in Fig. 4.10. The curve of pressure drop vs. air flow for a particular fan is known as its "fan curve".

The noise generated by a centrifugal fan depends strongly on where it is operating on its fan curve. Maximum efficiency of the fan, and minimum noise generation, occurs at a point just below the stall range. In most systems, the mechanical engineer will design the supply fan to operate in this region.

Acoustical treatments, particularly silencers, introduce additional pressure drop into the system, and the mechanical engineer will be particularly concerned about the pressure drop created by any additional devices recommended by the acoustic consultant. Too much additional pressure drop will move the fan into its stall range, which will significantly increase the SWL produced by the fan and also reduce its efficiency.

These issues obviously require liaison between the various consultants working on a project and are much easier to resolve in the early stages of the project. Acoustic design should ideally be among the first processes undertaken in designing a large mechanical ventilation system.

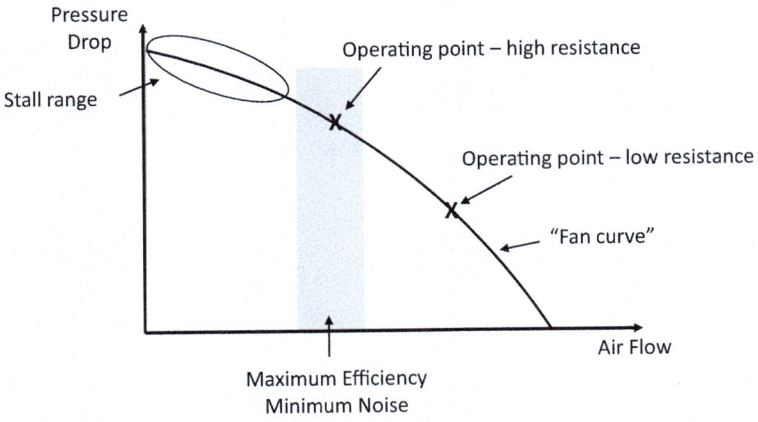

Fig. 4.10 Fan operating parameters

4.14 SPL in Occupied Spaces

Once the SWL injected into an occupied space has been determined for all source points, the SPL at a point in the room due to each source can be calculated as described in Chap. 2. This will depend on the room parameters as described there. Of course, the calculated SPL from all sources should be added logarithmically.

4.15 Other Sound Entering Ducts

If a duct is open to any space, sound from that space can enter the duct and travel to other spaces. This applies to sound from occupied spaces and also to sound from a plant room. It is common for return air ducts, in particular, to terminate in the plant room, allowing sound in the plant room to enter the open duct.

The sound power entering the duct in this way is calculated exactly as for the SWL passing through a partition, but with the R value of the partition being zero. For example, in a Sabine room (which applies to most plant rooms), Eq. (3.6) indicates that the SWL entering the duct will be:

$$\mathrm{SWL}_{\mathrm{entering}} = \mathrm{SPL}_{\mathrm{rev}} - 6 + 10\lg(S) \tag{4.3}$$

where $\mathrm{SPL}_{\mathrm{rev}}$ is the reverberant level in the room and S is the area of the open duct.

Once in the duct, this sound is attenuated and controlled in the ways described above.

Exercise 4.8

A fan has its return air inlet open to the plant room, producing sound power levels as follows:

Octave-band centre frequency, Hz						
63	125	250	500	1 K	2 K	4 K
82	84	78	72	68	65	60

The plant room has dimensions 4 × 3 × 3 m and has a reverberation time of 1s at all frequencies. The return air duct is open to the plant room, with dimensions 610 × 610 mm. (Direct sound from the fan to the duct opening can be ignored.)

What is the A-weighted SWL entering the return air duct from the plant room?

Appendix 1: Insertion Loss Values for Ducts

Unlined Ducts

The sound attenuation of an unlined duct refers to the difference between the SWL at the entrance and exit of the duct. Values of sound attenuation are tabulated in (ASHRAE 2019), in terms of dB reduction per metre of duct length. It is noted that attenuation for lengths greater than 3 m is not well documented.

Tabulated values of attenuation for unlined rectangular and circular ducts are reproduced in spreadsheet Supplementary file1. Figures 4.11 and 4.12 show examples of typical attenuations in terms of dB/m.

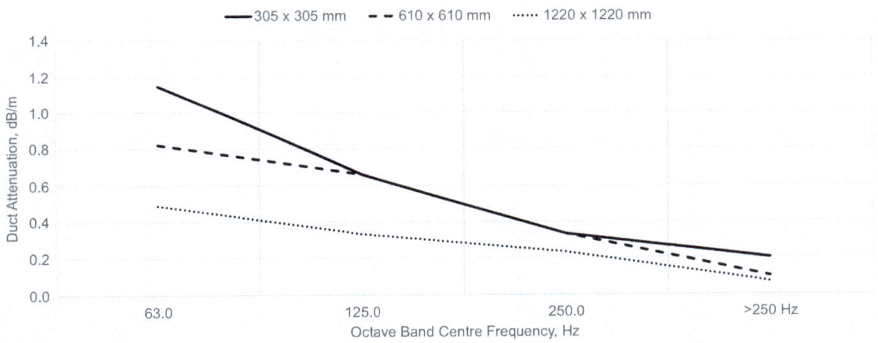

Fig. 4.11 Typical duct attenuation for unlined rectangular ducts

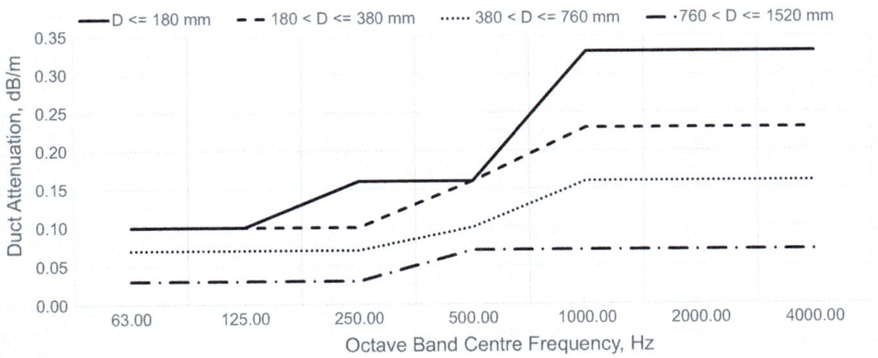

Fig. 4.12 Typical duct attenuation for unlined circular ducts

Lined Ducts

For lined ducts, attenuation is specified in terms of insertion loss—i.e. the difference between SWL at the exit with and without the lining—for a specified duct length. In this case, ASHRAE recommends the use of equations documented in (Reynolds and Schwob 2019a) and (Reynolds and Schwob 2019b) to calculate the insertion loss. These have been determined by regression analysis against a large set of data from lined ducts. The basic equations are as follows.

Rectangular ducts:

$$\mathrm{IL} = \left[a_1\left(\tfrac{P}{A}\right)^2 + a_2\left(\tfrac{P}{A}\right) + a_3\right] + L\left[b_1\left(\tfrac{P}{A}\right)^2 + b_2\left(\tfrac{P}{A}\right) + b_3\right]. \tag{4.4}$$

Circular ducts:

$$\mathrm{IL} = \left[c_1D^3 + c_2D^2 + c_3D + c_4\right] + L\left[d_1D^3 + d_2D^2 + d_3D + d_4\right], \tag{4.5}$$

in which

- $IL =$ insertion loss, dB,
- $L =$ length of lined duct, m,
- $P =$ perimeter of rectangular duct, mm,
- $A =$ open cross-sectional area rectangular of duct, mm^2,
- $D =$ diameter of circular duct, mm,
- $a_i, b_i, c_i, d_i =$ regression coefficients specified for each frequency band and depth of lining.

Notes:

1. The duct dimensions are *internal* dimensions—i.e. the lengths and areas refer to the area inside the lining.
2. The equations have been derived for:
   a. rectangular ducts with dimensions between 305 and 1220 mm and lengths between 0.9 m and 12 m,
   b. circular ducts with diameters between 305 and 1220 mm and lengths between 0.9 m and 9.1 m.

They should not be used outside this range.

The relevant regression coefficients are in the "Lined Rectangular Ducts" and "Lined Circular Ducts" pages of the Supplementary file1 spreadsheet. The same pages also include calculators allowing direct calculation of IL using Eqs. (4.4) and (4.5). Figures 4.13 and 4.14 show examples of typical insertion loss values.

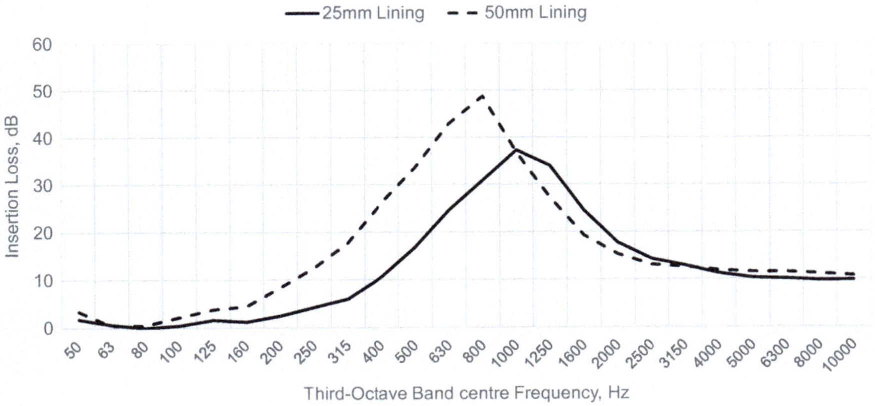

Fig. 4.13 Typical insertion loss for lined rectangular ducts—610 × 610 mm duct, 3 m length

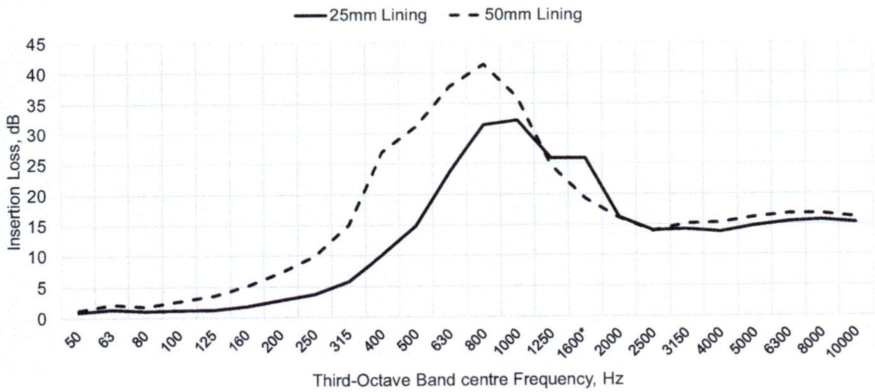

Fig. 4.14 Typical insertion loss for lined circular ducts—600 mm diameter, 3 m length

References

ASHRAE (2019) ASHRAE handbook—HVAC applications chapter 49: noise and vibration

Bradflo (2024) Round ceiling diffusors specification sheet, [Online]. Available: https://digitalassets.reecegroup.com.au/m/3273f709df27b0f6/original/Specification-Sheet-Bradflo-Aluminium-Round-Diffuser.pdf. Accessed Feb 2024

Chartererd Institution of Buildng Services Engineers (2005) Heating, ventilating, air conditioning and refrigeration: CIBSE Guide B

IAC Acoustics IAC duct silencer catalogue, IAC acoustics

Reynolds D, Schwob M (2019a) The effect of length on the insertion loss of fibreglass-lined sheet metal ducts—part 1: rectangular ducts, ASRAE research project RP-1408

Reynolds D, Schwob M (2019b) The effect of length on the insertion loss of fibreglass-lined sheet metal ducts—part 2: round ducts, ASHRAE research project RP-1408

Part II
Acoustic Design Principles

Chapter 5
Background Sound

Abstract This chapter describes the significance of background sound for various spaces; the ways in which background sound is measured and assessed; and criteria that should be specified in an acoustic design.

5.1 What Is Background Sound?

Background sound in any space can be defined as the components of the total acoustic environment that are relatively constant with time. Generally, background sounds would be characterised by changes of less than about $\pm$ 5 dBA in level over times of seconds to hours.

Background sound should therefore be distinguished from *ambient* sound, which generally means the entire acoustic environment including time-varying sounds. For example, background sound in an open-plan office may consist of noise from

- the air-conditioning system,
- computer fans,
- constant traffic noise entering through glazing.

Other sounds would form part of the ambient sound but would not be part of the background. These could include

- phone conversations,
- keyboards, shuffling papers, etc.

One way to distinguish background sound is that it will be present when the space is unoccupied.

Supplementary Information The online version contains supplementary material available at https://doi.org/10.1007/978-981-97-9380-8_5.

R. Bullen, *The Architecture of Sound*, Springer-AAS Acoustics Series,
https://doi.org/10.1007/978-981-97-9380-8_5

5.2 The Importance of Background Sound

Because of its quasi-steady nature, background sound is generally heard unconsciously—i.e. occupants of the space are not consciously aware that they are hearing the sound, unless they specifically listen for it. Nevertheless, it has significant implications for the occupants.

Studies such as (Wayne et al. 2002) have shown that the presence of background sound above certain levels can result in the release of additional stress hormones in people using the space. The effect appears to depend on (at least) the spectrum of the noise and the sensitivity of the occupants, which may differ for the same person in different environments. Beyond this, high levels of background noise have been associated with an increase in "arousal" (Garwon 1980), affecting various types of task performance. Again, the extent of the effects and the noise levels at which they occur are complex and not known accurately.

On the other hand, background sound can also be too quiet. In some spaces, this can result in an eerie feeling of silence, with any other noise becoming intrusive. An appropriate level of background noise can mask numerous other low-level noises, preventing them from being distracting.

5.3 Australian/New Zealand Standard 2107:2016

Australian/New Zealand Standard 2107:2016 (Standards Australia; Standards New Zealand 2016) presents a distillation of many years' practical experience in the design of background sound for numerous types of space. For each space listed, the standard gives a range of "design sound levels" in dBA. Background sound levels above the range risk causing stress and/or task performance effects, while levels below the range may result in insufficient masking. The latter effect is particularly important in spaces supporting several separate work areas, such as open-plan offices, classrooms and hospital wards.

Table 5.1 shows an example of recommendations from AS/NZS 2107 illustrating its use. On this basis, an executive office should have a background sound level not greater than 40 dBA L_{Aeq}. A high-quality executive office might be designed to 35 dBA, but levels below 35 dBA may result in distraction from other unmasked sounds.

Note that these recommendations do not specifically take account of requirements for speech intelligibility, in particular where electronic transmission is involved such as for video conferencing. This may put additional restrictions on background sound levels, and these are described in Chap. 9

Other standards provide acoustic guidance for certain specific types of space, as one part of general recommendations for that space. Examples include ISO 23591:2021 for music rehearsal rooms (International Standards Organisation 2021a), or ISO 22956:2021 for healthcare organisations (International Standards Organisation 2021b). In general, this advice is consistent with AS/NZS 2107.

Table 5.1 Recommended design sound levels for “Office Buildings”, from AS/NZS 2107

Type of occupancy/activity	Design sound level ($L_{Aeq,t}$) range
Board and conference rooms	30 to 40
Cafeterias	45 to 50
Call centres	40 to 45
Corridors and lobbies	45 to 50
Executive office	35 to 40
General office areas	40 to 45
Meeting room (small)	40 to 45
Open plan office	40 to 45
Public spaces	40 to 50
Quiet rooms	40 to 45
Reception areas	40 to 45
Rest rooms and break-out spaces	40 to 45
Toilets	45 to 55
Undercover carparks	< 65
Video/audio conference rooms	30 to 40

5.4 Other Background Sound Descriptors

As indicated in AS/NZS 2107 Appendix D, in some cases specifying the A-weighted SPL may not give sufficient control over the background sound. This may be because its spectral content is particularly unusual and potentially annoying or because the spectrum is not suitable to provide masking.

In these cases, more complex descriptors can be used. Various descriptors have been formulated and used historically, but the descriptor referred to in AS/NZS 2107 is known as Room Criterion (RC) Mark II. This is defined in (ASHRAE 2019) and summarised in Appendix 1 to this chapter. It allows identification of components of the background sound described as “low-frequency rumble”, “mid-frequency roar” and “high-frequency hiss” which may render the sound more annoying or less useful for masking than a “blander” spectrum.

Another common spectral descriptor of background sound is the noise rating (NR). This is also summarised in Appendix 1. Sometimes, it will be specified that the octave-band background SPL should match the values in a particular “NR curve”. These curves are given in the Supplementary file1 spreadsheet.

AS/NZS 2107 also provides a method for adjusting for the perceived additional impact of pure tones that may be present in the background sound. This is described in Appendix D of the standard and in Appendix 2 to this chapter.

Table 5.2 Recommended maximum octave-band SPL values where the recommended A-weighted SPL is 30 or 40 dBA L_{Aeq}

Octave-band centre frequency, Hz	31.5	63	125	250	500	1 K	2 K	4 K	8 K
Octave-band SPL, dB for a design level of 30 dBA	70	52	42	34	29	25	22	20	18
Octave-band SPL, dB for a design level of 40 dBA	80	62	52	44	39	35	32	30	28

For other A-weighted levels, the octave-band values can be increased or decreased equivalently

As an alternative to providing specifications in terms of other descriptors, the standard also provides recommended maximum octave-band SPL values which can be specified *in addition to* the A-weighted criteria described above. The octave-band criteria are shown in Table 5.2 for A-weighted criteria of 30 and 40 dBA and can be adjusted up or down for other A-weighted criterion levels.

This form of specification can in principle be used for any critical space, to provide greater control over the shape of the background spectrum. For example, in a library designed to a background level of 40 dBA the specification could also include recommended octave-band sound levels as shown in Table 5.2. However, in AS/NZS 2107, this form of specification is recommended only for studios, drama theatres and cinemas and its use for other spaces would be unusual.

Exercise 5.1

You are recommending design sound levels for a new hospital. The client wants the "highest standard of acoustic control".

1. What would you recommend as the A-weighted level of background sound in the patients' wards?
2. There is concern that background sound in the wards might have annoying characteristics, and to ensure these are minimised the client wants controls across the entire background noise spectrum. What would you set as maximum SPL values for each octave band?

5.5 Measuring Background Sound

In Australia, background sound levels are usually measured simply in terms of their L_{Aeq} value for comparison with criteria such as those in Table 5.1. However, care must be exercised as the intention is to measure the L_{Aeq} of the *background sounds only*. For example, in a working office there will be noise from voices, phones, etc., which do not form part of the background. Noise from people in an adjacent room or a corridor must also be excluded. The simplest way to ensure this is to conduct measurements when the space is ready for occupancy but unoccupied. Where this is not possible, it is sometimes necessary to estimate the background sound level from the lowest measurable level in the presence of other time-varying sounds.

5.6 Controlling Background Sound

In most spaces, there may be up to four sources of background sound:

- constant sound entering the space from outside, particularly traffic noise,
- constant sound from devices within the space, such as computer fans or room air-conditioners,
- sound from large-scale mechanical services, notably whole-of-building air-conditioning systems as described in Chap. 4,
- artificially introduced sound ("masking sound").

The specified background SPL values in AS/NZS 2107 represent the total SPL from all background sounds. For example, if both air-conditioning noise and traffic noise contribute to the background in a space, and the recommended maximum SPL from AS/NZS 2107 is 45 dBA, one option would be to set maximum levels of 42 dBA for each source.

The level of sound entering the space from outside is controlled by the design of partitions and control of flanking transmission and is discussed in detail in Chap. 6.

For devices within the space, control would be through device selection or, if necessary, enclosing the device. The sound pressure level due to a device or enclosure emitting a known sound power level can be calculated based on the principles described in Chap. 3.

In relation to sound produced by mechanical services, the major source of background sound in large buildings has traditionally been an air-conditioning system servicing the entire building. Prediction of sound levels from these systems is described in Chap. 4.

More recently, the use of passive ventilation and other more sustainable options has resulted in a decline in the use of large air-conditioning systems (see Chap. 10). This has led to increased use of artificially introduced "masking" sound in some spaces. This is discussed in Chap. 10.

Exercise 5.2

Suppose criteria have been set as in Exercise 5.1. Look at the Supplementary file1 spreadsheet under "Regenerated Noise" to find the regenerated SWL from diffusers. If

a. the only source of background SPL in the room is a diffuser with diameter 200 mm at 2 m from the listener,
b. the room is a Schultz room with volume 150 m^3.

What is the highest duct velocity that could be used without exceeding the criteria?

Appendix 1: Additional Measures of Background Noise

The RC Mark II Criterion

The RC Mark II criterion represents an alternative to the simpler A-weighted criterion for assessing the suitability of background noise. It is described in ASHRAE (2019) as follows.

> The RC Mark II method has three parts: (1) a family of criterion curves (Fig. 5.1), (2) a procedure for determining the RC numerical rating and the noise spectral balance (quality), and (3) a procedure for estimating occupant satisfaction when the spectrum does not have the shape of an RC curve (quality assessment index) ….
>
> The rating is expressed as RC followed by a number and a letter (e.g., RC 35[N]). The number is the arithmetic average rounded to the nearest integer of sound pressure levels in the 500, 1000, and 2000 Hz octave bands (the main speech frequency region) and is known as the preferred speech interference level (PSIL). The letter is a qualitative descriptor that identifies the sound's perceived character: (N) for neutral, (LF) for low-frequency rumble, (MF) for midfrequency roar, and (HF) for high-frequency hiss. There are also two subcategories of the low-frequency descriptor: (LFB), denoting a moderate but perceptible degree of sound-induced ceiling/wall vibration, and (LFA), denoting a noticeable degree of sound-induced vibration.

The procedure for determining the RC Mark II rating for background noise that has been measured or calculated in octave bands is:

> Step 1. Obtain the arithmetic average of the sound levels in the principal speech frequency range represented by the levels in the 500, 1000, and 2000 Hz octave bands … The RC reference curve is chosen to be that which has the same value at 1000 Hz as the calculated average value.
>
> Step 2. Calculate the quality assessment index (QAI) … which measures the deviation of the spectrum under evaluation from the shape of the RC reference curve. Calculate the energy-averaged spectral deviations from the RC reference curve in each of three frequency groups: low (LF; 16 to 63 Hz), medium (MF; 125 to 500 Hz), and high (HF; 1000 to 4000 Hz). (A simple arithmetic average of these deviations is often adequate for most engineering purposes.) Eq. (5.1) gives the procedure for the LF region; repeat for the MF and HF regions by substituting the corresponding values at each frequency.
>
> $$\mathrm{LF} = 10lg\left(\left[10^{0.1\Delta L_{16}} + 10^{0.1\Delta L_{31.5}} + 10^{0.1L\Delta}\right]/3\right) \tag{5.1}$$
>
> The ΔL terms are the differences between the spectrum being evaluated and the RC reference curve in each frequency band. In this way, three specific spectral deviation factors, expressed in dB with either positive or negative values, are associated with the spectrum being rated. QAI is the range in dB between the highest and lowest values of the spectral deviation factors.
>
> If QAI $\leq$ 5 dB, the spectrum is assigned a neutral (N) rating. If QAI exceeds 5 dB, the sound quality descriptor of the RC rating is the letter designation of the frequency region of the deviation factor having the highest positive value.

The following example is given.

> ***Example 1*** The spectrum plotted in (Fig. 5.1) indicated by large dots is processed …. The arithmetic average of the sound levels in the 500, 1000, and 2000 Hz octave bands is 35 dB, so the RC 35 curve is selected as the reference for spectrum quality evaluation.

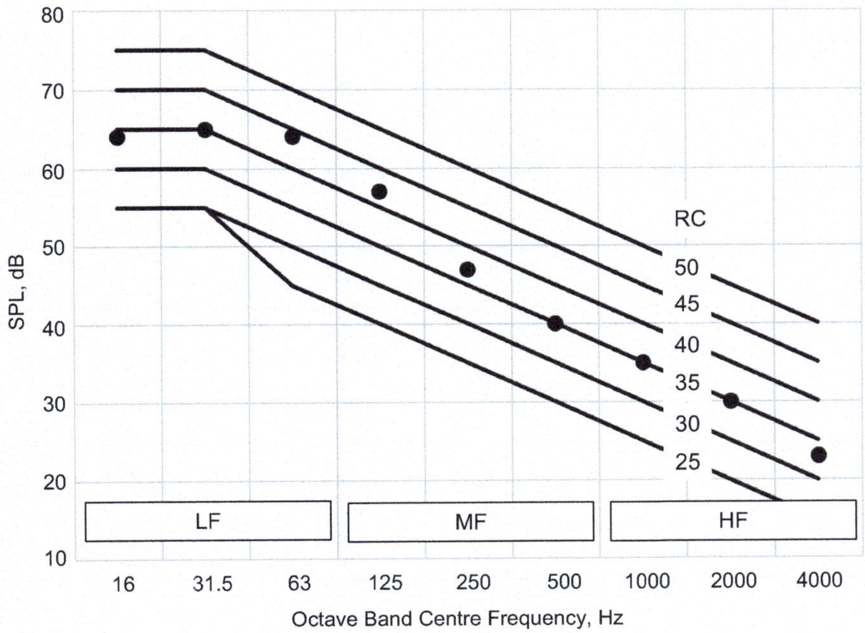

Fig. 5.1 RC MARK II curves with example spectrum (adapted from ASHRAE (2019))

The spectral deviation factors in the LF, MF, and HF regions are 6.6, 4.0, and −0.6 respectively, giving a QAI of 7.2. The maximum positive deviation factor occurs in the LF region and QAI exceeds 5; therefore, the rating of the spectrum is RC 35(LF). An average room occupant should perceive this spectrum as rumbly in character.

NR Curves

NR curves are used extensively in Europe to assess background noise in the way that the A-weighted noise level is used in Australia. The octave-band spectrum of the background noise is compared with the curves shown in Fig. 5.2 and the NR rating is assigned as the highest-value curve that is intersected by the spectrum.

In Australia, the actual NR curves are sometimes used as an indication of an "ideal" spectrum shape for background noise. Numeric values for the curves are given in the Excel spreadsheet for Chap. 5.

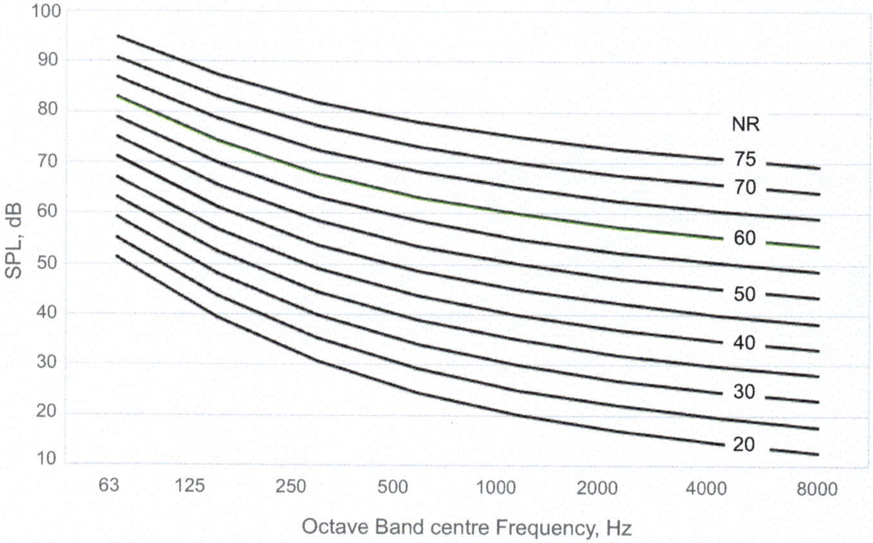

Fig. 5.2 NR curve values

Table 5.3 Level difference at which a correction for tonality is required under AS/NZS 2107:2016

Third-octave band centre frequency, Hz	Level difference, dB
25 to 125	15
160 to 400	8
500 to 10,000	5

Appendix 2: Accounting for Tonal Components in Background Sound

AS/NZS 2107:2016 contains a method of accounting for the presence of tonal components in measured background sound. This represents an adjustment of +5 dB to the measured A-weighted SPL, and its application is based on the difference between:

- the SPL in any third-octave band and,
- the arithmetic mean of the SPLs measured in the two adjacent bands.

The conditions under which the correction is applied are shown in Table 5.3. The correction is applied only once, even where the conditions in Table 5.3 are met for more than one frequency band.

A more robust, but more complex, method for detecting the presence of tonal components is set out in ISO PAS 20,065:2022, based on a narrow-band analysis of the signal. This leads to a value for "audibility" of the tone, ΔL, and ISO 1996-2:2017 then suggests a tonal adjustment based on the value of ΔL.

References

ASHRAE (2019) ASHRAE handbook—HVAC applications chapter 49: noise and vibration

Garwon V (1980) The effects of noise: a review. In: Proceedings of the human factors society—24th annual meeting

International Standards Organisation (2021a) DS/ISO 23591:2021 Acoustic quality criteria for music rehearsal room sand spaces. International Standards Organisation

International Standards Organisation (2021b) ISO 22956:2021 healthcare organization management—requirements for patient-centred staffing. International Standards Organisation

Standards Australia; Standards New Zealand (2016) AS/NZS 2107:2016 Acoustics—recommended design sound levels and reverberation times for building interiors

Wayne K, Rylander R, Hucklebridge F, Evans P, Clow A (2002) Low frequency noise enhances cortisol among noise sensitive subjects during work performance. Life Sci 70(7):745–758

Chapter 6
Intruding Sound—Airborne

Abstract This Chapter describes the significance of intruding sound in any space—that is, sound entering the space from outside. The level of audibility of intruding sound depends on masking by background sound, and requirements for audibility will depend on the nature of the space as well as the level of background sound. The chapter describes how to determine the level of intruding sound from other spaces and from the external environment; how to specify requirements for partitions to ensure that the level is appropriate; and common issues that limit achievement of good sound isolation. Associated spreadsheets provide ways to automate the calculations.

6.1 Intruding Sound and Its Control

Intruding sound is simply sound that enters a space after being produced by a source in another space, or in the external environment. This chapter considers "airborne" intruding sound—sound produced by a source in another space or in the external environment that enters through one or more partitions, or possibly through openings or gaps in those partitions.

The importance of intruding sound depends on the function of the space. For example, in a busy café, sound from the street may be welcomed as part of the atmosphere. However, in most spaces intruding sound can have negative impacts, including:

- distracting attention from activities within the space,
- compromising the atmosphere of the space,
- compromising speech privacy.

The required level of control for intruding sound depends on how important these effects are for the functioning of the space. The most critical level of control is *inaudibility*—i.e. sound entering the space should be either:

Supplementary Information The online version contains supplementary material available at https://doi.org/10.1007/978-981-97-9380-8_6.

R. Bullen, *The Architecture of Sound*, Springer-AAS Acoustics Series,
https://doi.org/10.1007/978-981-97-9380-8_6

- below the threshold of human hearing or,
- completely masked by background sound within the space.

Examples of cases where this would be appropriate are:

- when there is a requirement for total speech privacy in an adjoining space,
- a space where intense concentration is expected, e.g. a classical music venue,
- a space where maintenance of atmosphere is critical, e.g. a movie theatre.

Achieving inaudibility of intruding sound is often difficult and expensive and in many circumstances is not required. Where the requirement is to prevent distraction, it is often sufficient that the intruding sound be "buried" in the background sound, so it does not become a focus of attention. Achieving this goal can be significantly easier than achieving inaudibility.

Residential spaces are a particularly difficult case because requirements for acoustic control can depend on residents' expectations. For apartment and townhouse buildings, the Association of Australian Acoustical Consultants (AAAC) has approached this issue by producing a "star rating" system to define the level of sound intrusion achieved for transmission between different residences. This rating system is shown in Table 6.1.

Sound intrusion from sources in the external environment can result in more serious impacts for residents, including impacts on health and disturbance to sleep. In this case, criteria are set by environmental authorities.

Table 6.1 AAAC star rating system for ranking the sound intrusion achieved in apartments and townhouses (Association of Australasian Acoustical Consultants, 2017)

Type of noise source	Star rating for sound intrusion				
	2 star	3 star	4 star	5 star	6 star
Normal speech	Audible	Just audible	Not audible	Not audible	Not audible
Raised speed	Clearly audible	Audible	Just audible	Not audible	Not audible
Dinner party/laughter	Clearly audible	Audible	Just audible	Not audible	Not audible
Shouting	Clearly audible	Clearly audible	Audible	Just audible	Not audible
Small television/small entertainment system	Clearly audible	Clearly audible	Audible	Just audible	Not audible
Large television/large Hi-fi music system	Clearly audible	Clearly audible	Clearly audible	Audible	Just audible
DVD with surround sound	Clearly audible	Clearly audible	Clearly audible	Audible	Audible
Digital television with surround sound	Clearly audible	Clearly audible	Clearly audible	Audible	Audible

Table 6.2 Demonstration of speech in background noise (Open the associated Supplementary file to hear audio)

Difference between L_{Amax} of speech and L_{Aeq} of background
+5 dB [Link to Supplementary file3}
0 dB [Link to Supplementary file4}
−5 dB [Link to Supplementary file5}
−10 dB [Link to Supplementary file6}

In the absence of guidelines from authoritative sources, target levels for intruding sound need to be set according to the required level of audibility of the intruding sound in each space. This process is described in the following section.

6.2 Masking by Background Sound

Although in principle inaudibility could be achieved by reducing intruding sound levels to below the threshold of hearing, in practice there will be some level of background sound in any space, and audibility almost always depends on the extent to which this background sound masks the intruding sound.

Sound masking is the process in which one sound (the signal) is rendered inaudible by the presence of another, louder sound (the masker). The process of masking is complex and has been extensively studied. The degree to which one sound masks another depends on:

- the relative level of the masker and signal,
- the frequency difference between the masker and signal,
- whether the masker and/or the signal is tonal,
- the absolute level of the signal,
- time relationships between the masked and masking sounds.

(See (Painter & Spanias, 1977) for a full description).

Masking of intruding noise by background is considered in Appendix A. This detailed consideration leads to the following general "rule of thumb":

A typical intruding sound will be completely masked by a typical background sound if its maximum A-weighted level is about 10 dB below the A-weighted level of the background, measured as L_{Aeq}.

If the sound level is not at least 10 dBA below the background, an intruding sound will generally be audible, but will not necessarily be noticed by a listener. As its level rises, the intruding sound will move through the stages designated as "just audible", "audible" and "clearly audible" as shown in Table 6.1.

Once again as a "rule of thumb", the sound would generally be regarded as "clearly audible" if its maximum A-weighted SPL is about 5 dB above the A-weighted SPL of the background sound. Audio links in Table 6.2 demonstrate the audibility of speech at various levels with respect a background sound.

Exercise 6.1

(a) From Chap. 5, using guidance from AS/NZS 2107, what would be the maximum acceptable background noise level in a doctor's consulting room, in dBA?
(b) Assume that this maximum level is adopted. If another consulting room is adjacent, what should be the A-weighted maximum level of speech produced in one consulting room when heard in another?
(c) If a bus stop is outside, what would be maximum A-weighted level of noise from a bus, as heard in the consulting room, before it was clearly audible (and so potentially distracting).

6.3 Examples of Intruding Sound Calculations

Chapters 2 and 3 describe how to calculate the SPL of an intruding sound, given information on:

- the SWL of the source (or the SPL incident on the partition),
- properties of the spaces where the source and receiver are located,
- sound reduction values R for the partition.

The Supplementary file1 spreadsheet gives a number of examples of these calculations for various combinations of spaces, based on the tools described in Chapters 2 and 3. Remember that for these calculations sound level values for the source represent the *maximum* levels. For example, SWL values for a voice presented in the Supplementary file1 spreadsheet are L_{eq} values over a period of speech. Maximum values during that time can be expected to be about 5 dB higher at all frequencies.

Exercise 6.2

A person is standing close to a partition and on the other side is a 150 m^3 Schultz room, with a listener at 1.5 m from the talker.

Use the SWL spectrum for a shouted voice, from the Supplementary file1 spreadsheet, as the L_{eq} spectrum from their speech, directed at the wall, and assume that the maximum level is 5 dB higher at all frequencies. Adapt the "Source Close to Partition" sheet from the Supplementary file1 spreadsheet for this situation.

Detailed R values for various actual stud wall constructions are given in the publication here:

https://nrc-publications.canada.ca/eng/view/ft/?id=04ac8069-a5d2-4038-8787-da064b073e7f

(Look at pages 19 and after. "TL" values listed to the left of the page are R values.)

Use your adapted spreadsheet to find a wall construction that will keep the A-weighted SPL at the listener position below 35 dBA (but not too far below—do not over-design).

6.4 Flanking Transmission

The value of R is defined in terms of the sound power propagated into a receiving room *through a nominated partition.* In practice there may be many other paths by which sound from the source can enter the receiving room, including other partitions, gaps and weak points in the partition, and paths leading around the partition in various ways. All such paths are collectively described as "flanking transmission". Examples include:

- cross-talk through building services (such as ductwork) that passes over rooms and/or services more than one room,
- if the partition stops at the ceiling: transmission through the join between the partition head and the ceiling,
- if the partition stops just past the ceiling (i.e. penetrates the ceiling but does not extend full height to the underside of the roof/soffit above): transmission through the ceiling material and the ceiling cavity,
- transmission via penetrations in a partition for power points, light fittings, etc.,
- if there is a raised floor construction: transmission through the floor and underfloor area.

Measuring the value of R for a single partition requires flanking transmission to be negligible, which in practice almost always means that the measurement is performed in a laboratory. For these measurements, both the source and receiver rooms are designed to be Sabine spaces, so transmission is defined by Eq. (3.10), rearranged to give:

$$R = \mathrm{SPL}_{\mathrm{source}} - \mathrm{SPL}_{\mathrm{receiver}} + 10\,\lg(S) + 10\,\lg(T) - 10\,\lg(V) + 8, \qquad (6.1)$$

where T and V are respectively the reverberation time and volume of the receiving room and S is the area of the partition.

If a field measurement of R is performed between two rooms which can be assumed to be Sabine spaces, the measured value of reverberant SPL in the receiver room will include any flanking transmission in addition to transmission through the partition. If this value is substituted into Eq. (6.1), the result is known as the apparent sound reduction index R'. This has the effect of treating all sound transmission as if it was coming through the nominated partition, and so R' will be lower than R for the same partition design.

6.5 Sound Level Difference D

The sound pressure level difference D applies to sound transmission between two rooms, which may or may not be Sabine spaces, and is simply the (frequency-dependent) difference between the sound level measured in the source room and the resulting level measured in a receiving room. The measurement includes the effect of any flanking transmission.

The measured sound level in the receiving room, and hence D, will depend on the amount of absorption placed in the room and to partially correct for this effect D is usually converted to a "normalised" sound level difference

$$D_{nT} = D + 10\ \lg(T/T_0), \tag{6.2}$$

where T is the reverberation time of the receiving room and T_0 is a "standard" reverberation time—for residential spaces $T_0 = 0.5$ secs.

For transmission between two spaces, D_{nT} gives a direct measure of the level of intruding sound as heard in the space of interest—it is simply the sound level in the adjacent space minus D_{nT}, at each frequency. D_{nT} is also relatively easy to measure in the field, provided that the signal-to-noise ratio in the receiving room can be made greater than 10 dB.

For this reason, where control of intruding noise is critical, it is useful to specify a required sound level difference between spaces in terms of D_{nT} values rather than specifying the performance of partitions in terms of R values. This allows for a test to be undertaken to verify that the overall construction, including flanking transmission, will achieve the required level of control over intrusive noise, and puts the onus on the builder to achieve compliance.

Of course, at the design stage the final D_{nT} value needs to be inferred from the R values for individual partitions together with specified measures to control various forms of flanking transmission.

The above discussion refers to transmission between two rooms. International Standard ISO 16283:3 (International Standards Organisation, 2016) is specifically applicable to facades—i.e., sound entering a building from external sources. This standard specifies a similar unit called $D_{2\text{m},nT}$, representing the actual difference between an external sound level (measured at 2 m from a façade) and the resulting internal sound level. However, a value for this sound reduction performance is rarely specified in practice (at least in Australia).

6.6 Weighted Sound Reduction Measures

R, R' and D_{nT} are all frequency-dependent and are usually reported in third-octave bands between 100 Hz and 5 kHz. (Special low-frequency procedures are defined that can give values at 50, 63 and 80 Hz.) As noted in Chap. 3, they will generally be higher at higher frequencies, although there may be dips and other features in the plot of sound reduction versus frequency. It is useful to define a single number summarising the performance of the partition or of the building design in the case of D_{nT}.

The definition of these single-number values is given in Australian/New Zealand Standard AS/NZS ISO 717.1 (International Standards Organisation, 2020). All three use a procedure for adapting measured values to a reference curve and quoting the value of that curve at 500 Hz, resulting in "weighted" values of R, R' and D_{nT} referred to as R_w, $R_{w'}$ and $D_{nT,w}$, respectively.

While these single-number values are used widely, they do not necessarily provide a good rank ordering of the performance of partitions in reducing the level of sounds with different spectra, nor do they provide a good estimate of the reduction in the A weighted level of a sound source. AS/NZS ISO 717.1 introduces two correction factors: C and C_{tr}. These can be used to determine spectrum adaption terms which may be used to give a better estimate of the performance of the partition for different types of sources.

The C correction factor is used to assess the performance of the partition at reducing noise from relatively high-frequency sound such as talking, railway trains, road traffic at high speeds and nearby jet aircraft. The C_{tr} correction factor is used when the noise sources are primarily lower frequency sound such as urban road traffic, low speed rail traffic, distant jet aircraft and music with low-frequency content.

The procedure for calculating the spectrum adaptation terms is described in the standard and is based on calculating the reduction in A-weighted SPL from sources with two different spectra (with standardised source and receiver spaces).

The resultant values C and C_{tr} are generally negative numbers and are quoted in brackets after the weighted value. For example, one could write $R_w(C; C_{tr}) = 41(-1; -5)$ or $D_{nT,w}(C; C_{tr}) = 35(0; -8)$. An approximation to the difference in A-weighted SPL between two standard rooms is then the weighted value plus the correction appropriate for the source spectrum.

6.7 Examples

To illustrate the usefulness and also the limitations of the procedure described in Sect 6.6, we consider two recorded voice samples, one female and one male, sourced from (Christian-Albrechts-Universität zu Kiel, n.d.). The spectra of the two samples are shown in Fig. 6.1 and compared with the "average" spectrum for a loud voice from the Supplementary file1 spreadsheet, with all spectra referenced to a level of 75 dBA.

We can calculate the SPL from these sources in another room, after passing through one of two different walls:

- 13 mm plasterboard stud wall with no insulation—$R_w(C; C_{tr}) = 38(-2; -7)$,
- 16 mm plasterboard stud wall with insulation—$R_w(C; C_{tr}) = 44(-2; -6)$.

Based on these R_w and C values, voices at 75 dBA should result in A-weighted sound levels of about $75 - (38 - 2) = 39$ dBA for the first wall and $75 - (44 - 2) = 33$ dBA for the second.

Results of detailed calculations, using the third-octave spectrum values for the voices and full R values for the walls, are shown in Table 6.3. The calculations assume:

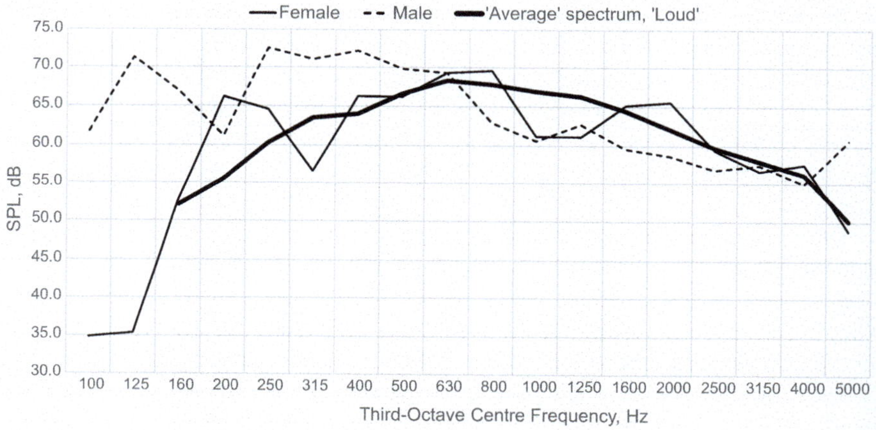

Fig. 6.1 Spectra from recorded female and male voice samples and the "average" spectrum for a "loud" voice, all referenced to a total level of 75 dBA

Table 6.3 Levels of speech at 75 dBA L_{Amax} heard through a plasterboard wall. (Open the associated Supplementary file to hear audio with 35 dBA background noise in the receiving room)

Wall system	Voice type	Maximum SPL in receiving room, dBA	
		Detailed calculation	Using $R_w + C$
13 mm plasterboard, no insulation	Female	35 [Link to Supplementary file7}	39
	Male	40 [Link to Supplementary file8}	39
16 mm plasterboard, 50 mm insulation	Female	30 [Link to Supplementary file9}	33
	Male	33 [Link to Supplementary file10}	33

- two adjacent Sabine rooms,
- reverberant level 75 dBA L_{Amax} in the source room,
- partition area 10 m^2,
- receiver room 70 m^3 with reverberation time 0.9 s at all frequencies.

Table 6.3 has links to audio files demonstrating the sound of the female and male speech as heard through these two wall systems, in the presence of background sound at 35 dBA in the receiving space. Imagine you are in a work environment and you hear these voices through a wall. Would you be disturbed?

All these calculations, of course, neglect the effect of flanking transmission on the received sound level. In the absence of specific controls, a "rule of thumb" is that flanking transmission can increase the received SPL by up to 5 dBA.

6.8 Sound Reduction Properties of Single Panels

For a single uniform panel, a chart of R versus frequency can be divided into distinct regions, as shown in Fig. 6.2 (based on (Bies & Hansen, 2009), Sect. 7.2.4).

Below the "first panel resonance" frequency, R is determined by the stiffness of the panel and is relatively constant with frequency. The value of this resonance frequency depends on the size of the panel and properties of the material, but is generally lower than 100 Hz for standard building materials. An explicit formula can be found in (Bies & Hansen, 2009).

In a region after this, the value of R is determined by the effect of subsequent resonances and anti-resonances, with behaviour that cannot be predicted by a simple schema.

At a certain frequency, the value of R meets a line defined by the "mass law":

$$R = 20 \lg(fm) - 48, \tag{6.3}$$

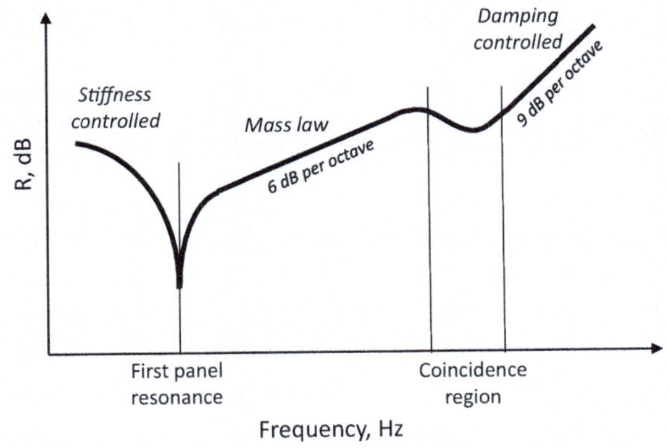

Fig. 6.2 Schematic chart of R for a single uniform panel, adapted from (Bies & Hansen, 2009)

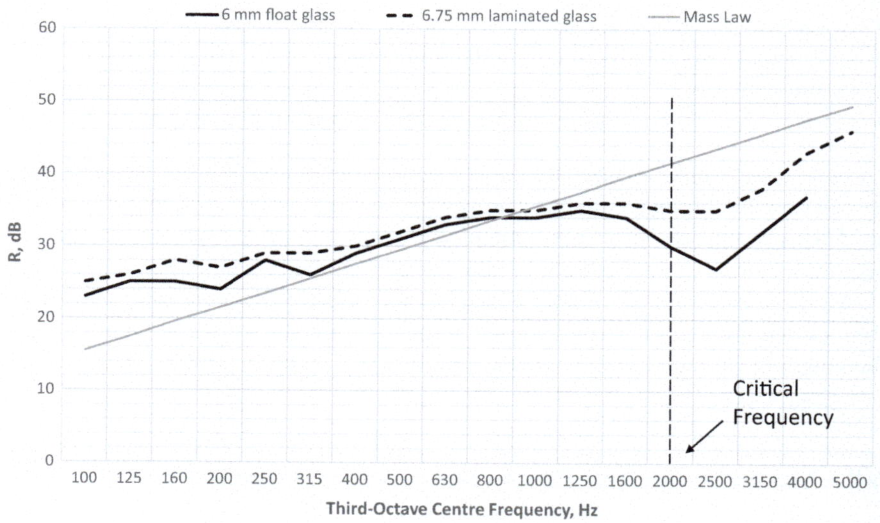

Fig. 6.3 Measured R values for 6 mm glass with and without laminate for additional damping (data from (Viracon acoustic performance data tables, n.d.))

where f is frequency and m is the mass per unit area, or "area density", of the panel. In this region, the value of R increases by 6 dB per octave and 6 dB per doubling of m.

This continues until the "critical frequency", which is given in (Bies & Hansen, 2009) as

$$f_c = \frac{0.55c^2}{c_L h}, \tag{6.4}$$

where c is the speed of sound in air, c_L is the longitudinal wave speed in the material and h is the thickness of the panel.

At the "critical frequency", the wavelength of the sound equals the bending wavelength of the material, and bending waves can be generated if the sound arrives at grazing incidence to the panel. This results in increased sound transmission. At higher frequencies, the bending wavelength matches the projected wavelength for sound at other angles of incidence, resulting in an area of reduced sound transmission known as the "coincidence region" or "coincidence dip".

The "depth" of the coincidence dip is controlled by damping in the material. For frequencies above the coincidence region, the value of R rises at approximately 9 dB per octave until it meets an extension of the "mass law" curve.

Figure 6.3 illustrates the effect of the coincidence dip, which occurs at approximately 2 kHz for 6mm glass, and the effect of additional damping in the material due to lamination, on measured R values (data from (Viracon acoustic performance data tables, n.d.)).

Exercise 6.3

The "Source in External Environment" sheet of the Supplementary file1 spreadsheet calculates the SPL from a heavy vehicle through a window with 6.75 mm laminated glass.

(a) What would be the increase in noise level if the laminated glass were replaced with 6 mm float glass? R values for each of these types of glass are in the "Data for exercise 6.3" sheet.
(b) The R values for the 6.75 mm laminated glass give $R_w(C; C_{tr}) = 35(-1; -2)$. The A-weighted level of the vehicle outside is 78.2 dBA. If you used only $R_w(C; C_{tr})$, what would you calculate for the internal noise level through the glass (to the nearest dB)? Look at how this compares with the calculated levels in a Sabine and a Schultz space.
(c) The detailed calculations are through a small window. If the 1.8 m × 1.5 m dimensions for the partition were replaced by a fully glazed façade, say 6 m × 2.8 m, with 6.75 mm laminated glass, what is the calculated internal noise level now?

6.9 Sound Reduction Properties of Double Panels

Double partitions such as framed and sheeted cavity walls or double-glazed windows are commonly used to achieve a higher sound reduction than single partitions. This higher sound reduction can be achieved with less overall mass—an important factor to be considered in multistorey buildings. However, due to coupling of the vibration from one panel to the other (sometimes by air), the sound reduction is not simply the sum of the R values for each partition, unless the panels are several metres apart.

Techniques to increase the value of R for a double panel partition include:

- increasing the mass of the panels—either by selecting a different material with a higher surface density, increasing the thickness of the panel or increasing the number of panels,
- using materials with increased damping, generally by using composite materials for the panels. This applies particularly to glazing, where the use of a laminated interlayer can significantly increase the performance of the panel,
- minimising the direct connections from one side to the other. Many wall systems incorporate specially designed framing elements that reduce the transmission of vibration from one leaf to the other while maintaining the structural integrity of the partition (e.g. separate studs, staggered studs, and resilient mounts),
- increasing the width of the cavity, as this reduces the stiffness of the air in the cavity and lowers the frequency of the coupling resonance,

- including sound absorbing material in the cavity, which can also reduce the effect of the cavity resonances. For double glazing, sound absorbing material can be installed in the reveals to maintain transparency,
- using panels of different thickness or mass, which helps to ensure that dips in the performance of each panel due to resonances and the coincidence effect do not occur at identical frequencies,
- offsetting the joints between layers when multiple layers are used on one or both sides.

Manufacturers of products designed for acoustic performance will have data for numerous designs, generally presented in terms of R_w, $R_w + C$ and $R_w + C_{tr}$ values. For example, Fig. 6.4 shows R_w and $R_w + C_{tr}$ values for a sample of stud wall constructions, all using plasterboard on both sides with 50 mm absorbing material in the cavity, based on data in (CSR, n.d.).

To calculate the level of intruding sound in more detail, R values are required at each frequency. These are sometimes available from manufacturers' data, but otherwise, for complex panel constructions, their calculation requires significant work. Relevant formulae and schemas can be found in (Bies & Hansen, 2009).

Alternatively, Insul (Marshall Day Acoustics Ltd, n.d.) is a software tool that can be used to predict the sound insulation properties provided by complex wall, floor and ceiling constructions, including transmission loss, impact sound and rain noise.

6.10 Design Guidelines

As outlined in Sect. 6.1, the aim of intrusive noise control is to achieve a level of audibility consistent with the requirements of the space, which may range from "not audible" to "clearly audible". The level of audibility will depend on the sound level of the intruding sound and the sound level of the background sound in the receiving space. Design sound levels for background sound will be set using principles described in Chap. 5, while the level of intruding sound is controlled using the principles described above.

Where control of intrusive sound is critical, control should be based on

- a full third-octave analysis of the sound from the source,
- R values for the partition(s),
- consideration of the types of source and receiving spaces (Sabine or Schultz),
- detailed consideration of possible flanking transmission paths.

In less critical cases, the control can be based on

- the A-weighted source level,
- $R_w + C$ or $R_w + C_{tr}$ values for the partition(s),
- an assumed 5 dB penalty to account for flanking transmission.

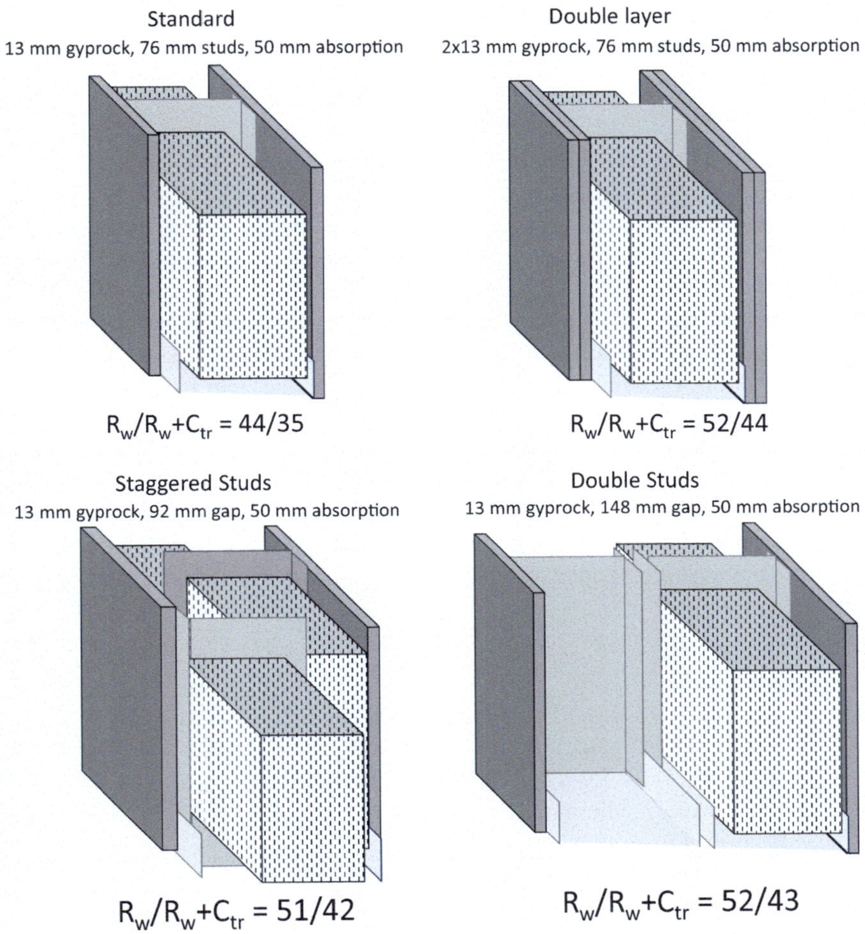

Fig. 6.4 Acoustic performance of a sample of wall constructions—data from (CSR)

However, it should be borne in mind that calculations based on single-number ratings for partitions can give results that are incorrect by several dB, and the calculated noise levels may be either too high or too low.

Although calculations will necessarily be in terms of R, $R_w + C$ or $R_w + C_{tr}$ values, specifications for buildings should, if possible, be given in terms of D_{nT}, $D_{nT,w} + C$ or $D_{nT,w} + C_{tr}$ values, as these values can be checked once the building is constructed.

Ultimately, the audibility of intruding noise is a decision for the acoustic designer and should be made in consultation with the end user of the space. However, for some standard spaces, guidelines produced by authorities or professional bodies may be used. The Association of Australasian Acoustical Consultants produces guidelines for intruding noise in:

- apartments and townhouses,
- commercial buildings,
- educational facilities,
- healthcare facilities.

These are available from https://aaac.org.au/Guidelines-&-Downloads.

For apartments and townhouses, the guidelines include recommended $D_{nT,w}+C_{tr}$ values for partitions between occupancies, or D_{nTw} values for doors, to achieve the "star rating" values described in Table 6.1. The recommended values are shown in Table 6.4.

For intrusion by external noise, the AAAC guideline recommends absolute noise levels to be achieved in bedrooms and other habitable rooms, with the design and rating of partitions (generally glazing) being left to the designer based on the level of external sound.

In the AAAC guidelines for commercial, educational and healthcare facilities, recommendations for control of internal noise intrusion are in terms of D_w values of partitions, which can be roughly equated with $D_{nT,w} + C$. Required values are determined on the basis of the usage of the space and of the adjacent space, which determines both the importance of sound isolation and the likely background sound level.

Exercise 6.4

(a) From the AAAC guidelines for educational facilities, what is the recommended D_w value for a wall between a space with "high activity noise" and one with "medium noise tolerance"?

(b) Assuming D_w is approximately $D_{nT,w} + C$, how would you specify a required performance for this wall in terms of R values from manufacturers' data?

Table 6.4 AAAC guidelines for partitions between occupancies in apartments and townhouses (Association of Australasian Acoustical Consultants, 2017)

Intertenancy activities	Star rating for sound intrusion				
	2 star	3 star	4 star	5 star	6 star
Between separate tenancies $D_{nT,w} + C_{tr}$	35	40	45	50	55
Between a lobby/corridor and a bedroom $D_{nT,w} + C_{tr}$	30	40	40	45	50
Between a lobby/corridor and a living area $D_{nT,w} + C_{tr}$	25	40	40	40	45
Corridor, foyer to living space via door(s) $D_{nT,w}$	20	25	30	35	40

6.11 Practical Considerations

In order to achieve the performance predicted using the procedures outlined above, careful attention to construction detail is essential. The potential faults in construction that can affect the overall acoustic performance include:

- poor sealing at partition junctions, such as the head join to the ceiling material and infill panel joins to the facade, etc.,
- not sealing around penetrations for mechanical or other building services in full-height walls,
- mechanical linkages between partition leaves that are designed to be separate. For example, debris falling into double stud wall cavities, studs that are not installed with a sufficient gap between internal faces, mortar for a brick veneer wall forming a bridge between the brick wall and stud work, etc.,
- continuous lightweight elements without a structural break that span across two spaces—e.g. a continuous ceiling between two spaces will limit the achievable separation between two spaces,
- poor fitting door seals that do not achieve an air-tight compressive seal.

Site inspections during the construction phase are essential to ensure that works are undertaken in accordance with recommendations and that the intent of the acoustic design is being maintained.

Appendix 1: Masking by Background Sound

Masking of a Tone by Noise

Masking of sound by other sound has been extensively studied, particularly in the context of designing audio codecs where the intention is that only audible sound should be transmitted. In the context of building acoustics, the important question is the extent to which intruding sound, which may or may not be tonal, is masked by background sound, which will always be non-tonal. Because masking of tonal sound is more difficult, attention will be focussed on masking of a tonal sound by a non-tonal masker. Also, because background sound can be assumed to be relatively steady, effects related to temporal changes in the masker are not relevant.

The process of auditory masking is described in detail in (Painter and Spanias 1997).

For any intruding sound in the presence of a masker, the "masking threshold" is the level of the intruding sound, in dB, below which it is inaudible. This threshold depends on the levels of the intruding sound and the masker, as well as their relative frequencies.

Perceptually, frequency is described on a "bark" scale, where the bark value z is related to the frequency f by

$$z = 13\tan^{-1}\left(\frac{f}{1316}\right) + 3.5\tan^{-1}\left[\left(\frac{f}{7500}\right)^2\right]. \tag{6.5}$$

A difference of one bark is described as a "critical bandwidth". This bandwidth is approximately 100 Hz for frequencies below 500 Hz, and 20% of the centre frequency at higher frequencies. Studies of masking of a tone by noise focus on determining the masking threshold for a tone at frequency z_1 barks by noise with a bandwidth of 1 bark centred on a frequency of z_2 barks. This masking threshold is given by (Painter & Spanias, 1997)

$$\mathrm{MT}(z_1) = \mathrm{SPL}_{\mathrm{noise}} - 0.175z_2 - 2.025 + \mathrm{SF}(z_1 - z_2), \tag{6.6}$$

where $\mathrm{SPL}_{\mathrm{noise}}$ is the level of masking noise in the critical bandwidth and SF is a "spreading function", equal to zero when $z_1 = z_2$. When z_1 is not equal to z_2, SF is given by

$$\mathrm{SF}(\Delta z) = \begin{cases} 17\Delta z - 0.4\mathrm{SPL}_{\mathrm{noise}} + 11 & \Delta z < -1 \\ (0.4\mathrm{SPL}_{\mathrm{noise}} + 6)\Delta z & -1 \leq \Delta z < 0 \\ -17\Delta z & 0 \leq \Delta z < 1 \\ (0.15\mathrm{SPL}_{\mathrm{noise}} - 17)\Delta z - 0.15\mathrm{SPL}_{\mathrm{noise}} & \Delta z \geq 1 \end{cases} \tag{6.7}$$

where $\Delta z = z_1 - z_2$.

This function is demonstrated in Fig. 6.5.

From Fig. 6.5, it can be seen that:

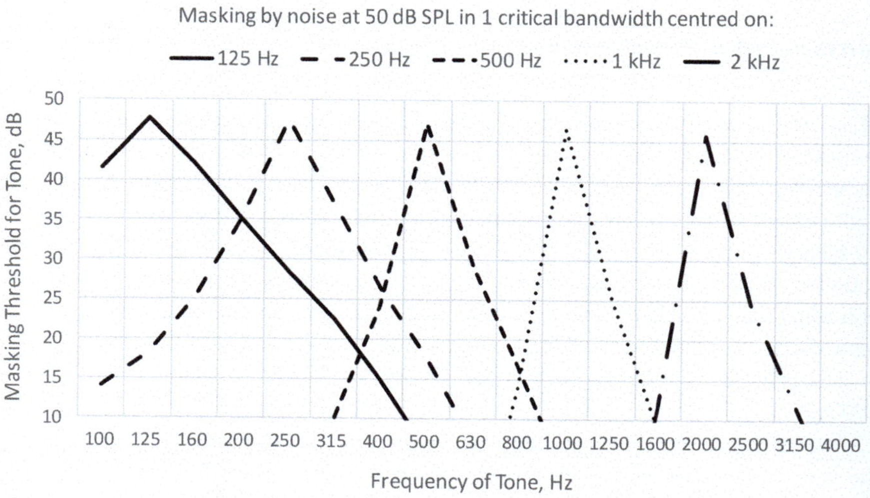

Fig. 6.5 Masking threshold for a tone masked by noise with a level of 50 dB and bandwidth of 1 critical bandwidth centred on the frequencies shown

- when a tone is masked by noise centred on the frequency of the tone, the masking threshold is a few dB below the level of the noise,
- when the frequency of the tone differs from the centre frequency of the noise, the masking threshold reduces quickly, particularly at high frequencies.

Application to Masking of Intruding Sound

Intruding sound may be tonal or non-tonal (and is generally a combination of both), but to provide a conservative estimate of masking, we can assume that in each frequency band, the intruding sound consists of a tone at the centre frequency of that band. Masking will arise mostly from background sound in the same frequency band. For the masking to be complete, the intruding sound must be below the masking threshold in each band.

Analysis by bark bands is often not available, but if both the intruding sound and the masking sound are measured in third-octave bands, this gives a reasonable approximation. The masking threshold for tonal noise is calculated for the centre frequency of each third-octave band, and this is compared with the level of intruding noise in that band.

Figure 6.6 shows an example of masking by background noise. The light solid line shows a background noise spectrum following the NR40 curve with the level set at 50 dBA. For each band, the dashed line shows the masking threshold for intruding tonal noise in that band that would result from the background noise in that band. The heavy solid line shows a typical spectrum for female speech, also at 50 dBA.

The speech spectrum exceeds the masking threshold by up to 10 dB, which implies that the spectrum needs to be reduced by 10 dB for the speech to be completely masked. Of course, the level of speech varies with time, so this criterion must apply to the maximum level. To be completely masked, the L_{Amax} level of the speech should not exceed 40 dBA.

This represents a general "rule of thumb"—a typical intruding sound will be completely masked by a typical background sound if its maximum A-weighted level is about 10 dB below the A-weighted level of the background, measured as L_{Aeq}.

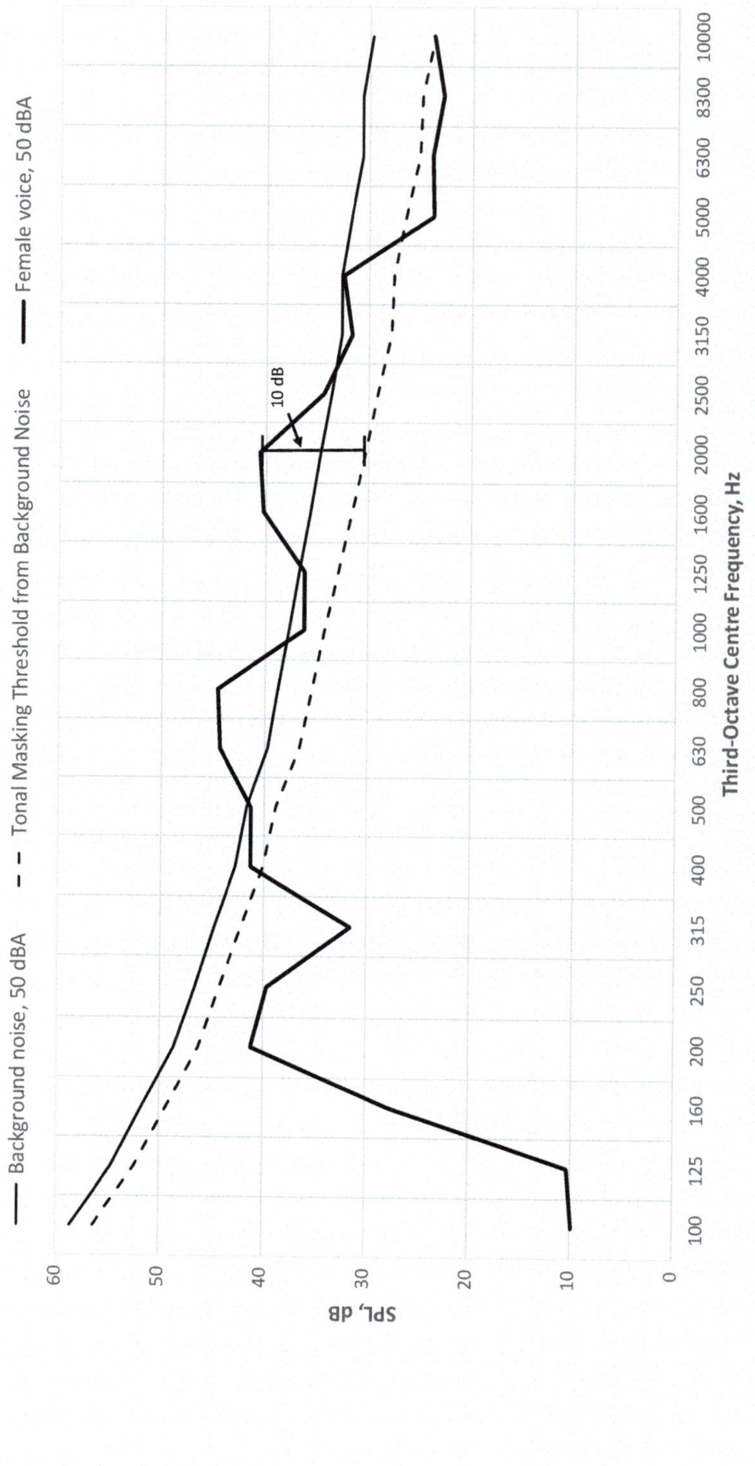

Fig. 6.6 Background noise and female voice spectra, both at 50 dBA, and masking level for tonal noise at each third octave

References

Association of Australasian Acoustical Consultants (2017) Guideline for apartment and townhouse acoustic rating, V 1.0

Bies D, Hansen C (2009) Enginering noise control, 4th edn. Spon Press

Christian-Albrechts-Universität zu Kiel, Faculty of Engineering, institute for electrical engineering and information engineering, digital signal processing and system theory, Lombard speech examples, [Online]. Available: https://dss-kiel.de/index.php/media-center/audio-examples. Accessed March 2024

CSR. The gyprock red book, [Online]. Available: https://www.gyprock.com.au/resources/redbook

International Standards Organisation (2016) ISO 16283-3:2016 Acoustics—field measurement of sound insulation in buildings and of building elements. Part 3: facade sound insulation. International Standards Organisation

International Standards Organisation (2020) ISO 717-1:2020 Acoustics-rating of sound insulation in buildings and of building elements. Part 1: airborne sound insulation. International Standards Organisation

Marshall Day Acoustics Ltd, Insul, [Online]. Available: https://www.insul.co.nz

Painter T, Spanias S (1977) A review of algorithms for digital coding of audio signals. In: 13th annual conference on digital processing

Viracon acoustic performance data tables, Viracon, [Online]. Available: https://www.viracon.com/wp-content/uploads/2020/09/ViraconAcousticPerfDataTables.pdf. Accessed March 2024

Chapter 7
Intruding Sound—Structure-Borne

Abstract This chapter describes the entry of intruding sound through vibration of the building structure, including propagation of vibration through the building and generation of sound in the receiving space. Measurement and standards for vibration and structure-borne noise are described, as well as techniques for controlling vibration. Associated spreadsheets provide ways to automate the calculations.

7.1 Vibration and Intruding Sound

Another way in which sound can intrude into a space is through vibration of the building elements—floor, walls and/or ceiling—that surround the space. The vibration could be generated at another point in the building, or in the external environment, and propagates through the building structure. Sound in the receiving space is generated directly by vibration of the surrounding building elements. This is known as "structure-borne sound".

The most common example of structure-borne sound is from the impact of hard-heeled shoes on the floor resulting in sound in a space below. However, there are many other examples, including:

- dropping weights in a gym,
- jumping in an aerobics class,
- driving over speed bumps in an above-ground carpark,
- water movement in an upper-floor swimming pool,
- mechanical plant including fans, chillers, dishwashers, etc.,
- "water hammer" from plumbing fixtures,
- construction equipment, either within or outside the building,
- train movements close to the building.

Supplementary Information The online version contains supplementary material available at https://doi.org/10.1007/978-981-97-9380-8_7.

R. Bullen, *The Architecture of Sound*, Springer-AAS Acoustics Series,
https://doi.org/10.1007/978-981-97-9380-8_7

Some of these vibration sources are primarily high frequency (e.g. impacts from shoes), while others are primarily low frequency (e.g. dropping weights).

Once generated in the space, structure-borne sound can be assessed in the same way as airborne sound—i.e., the maximum SPL of the intruding sound is compared with the background sound in the space to determine its audibility, and a design decision made as to whether that level is appropriate. However, estimation of the likely SPL and design of mitigation measures are more complicated than for airborne noise.

7.2 Vibration Velocity

As described in detail in Appendix 1, the most important quantity in describing vibration in a building is the mean-square vibration velocity $\langle v^2 \rangle$ in each building element. This is expressed in decibel terms as the velocity level

$$L_V = 10\ \lg\left(\frac{\langle v^2 \rangle}{v_0^2}\right) \tag{7.1}$$

with $v_0 = 10^{-9}$ m/s.

The vibration velocity level in a building element is related to the force producing the vibration and is also directly related to the resulting SPL in the space.

The relationship between an applied force and the resulting vibration velocity is complicated, as described in Appendix 1. In particular, at low frequencies there will be resonances associated with vibrational modes of the element which can increase vibration levels at certain frequencies by up to 10 dB.

Transmission of vibration to other building elements introduces further complexities. Between any two elements, there will be a "coupling loss factor" that depends on the material properties of the elements and how they are connected. It is also possible that the second element will experience resonances at certain frequencies.

In practice, calculation of vibration levels throughout a building structure from a known force input requires a computer model—generally of a type known as a "finite element model". However, for high frequencies and/or large building elements, approximations can be used as described in Appendix 1.

Alternatively, if the building structure exists, one can use a standard vibration source and measure either vibration levels, or the resulting sound pressure levels in spaces, directly.

Fig. 7.1 Standard tapping machine, without cover Image courtesy of Renzo Tonin & Associates

7.3 ISO Tapping Machine

A tapping machine, shown in Fig. 7.1, is a standard source of vibration defined in ISO 16283-2. It includes hammers, each weighing 0.5 kg and dropped from a height of 0.04 m, with ten impacts per second. This is intended to create a force roughly equivalent to walking with hard-heeled shoes.

As described in Appendix 1, at high frequencies a tapping machine will produce equal levels of vibrational energy in each octave or third-octave band.

7.4 Forms of Excitation

Three forms of vibration excitation should be distinguished.

- **Direct** excitation occurs when an impact, such as from a tapping machine, strikes a building element directly. In this case, the resulting vibration velocity depends strongly on the density and other properties of the material that is impacted. There is an approximation, valid at high frequencies, that allows vibration velocities to be predicted for a known force, based on bulk properties of the material. This is described in Appendix 1 and is implemented in the Supplementary file1 spreadsheet using the force input from a standard tapping machine on concrete. At lower frequencies, a full finite element model is required.

- **Indirect** excitation occurs when vibration created by a force on one building element is transferred to others. When the excitation is indirect, vibration velocities still depend on the properties of the affected element, but not to the same extent as for direct excitation. Transfer also depends, of course, on the type of coupling involved between the elements.
- **Velocity** excitation occurs when vibration is transferred to the building from the adjacent ground, for example in the case of rail vibration. In this case, the resulting vibration velocities will the same in all building elements that are in contact with the ground, independent of the properties of the element. Calculation of the input vibration levels from various sources is beyond the scope of this book.

7.5 Sound Radiation

The sound power radiated from a vibrating plate depends on the vibration velocity in the plate according to

$$\mathrm{SWL} = L_v + 10\,\mathrm{lg}(S) + 10\,\mathrm{lg}(\sigma) - 34, \tag{7.2}$$

(Bies and Hansen (2009)) where S is the area of the plate and σ is the "radiation efficiency" of the plate, described below. If the sound power is radiated into a Sabine space, Eq. (2.1) gives

$$\mathrm{SPL} = L_V + 10\,\mathrm{lg}(\sigma) + 10\,\mathrm{lg}\left(\tfrac{S}{V}\right) + 10\,\mathrm{lg}(T) - 20, \tag{7.3}$$

where V is the room volume and T is the reverberation time. If the radiating surface is a floor or ceiling, V/S is the ceiling height. Assuming a typical residential space with $V/S \sim 2.8$ m and $T \sim 0.5$s gives

$$SPL = L_v + 10lg(\sigma) - 27. \tag{7.4}$$

Similar calculations for a Schultz space give a result about 1 dB lower.

The radiation efficiency of a plate, σ, depends on the ratio of the vibration frequency to the *critical frequency* of the plate, as well as on other properties of the plate. This critical frequency is the same as the critical frequency f_c described in Sect. 6.8 and is given by Eq. (6.4).

The Supplementary file1 spreadsheet calculates values of 10 lg(σ) for various materials, using formulae from (Bies and Hansen 2009). Figure 7.2 shows a selection of the results.

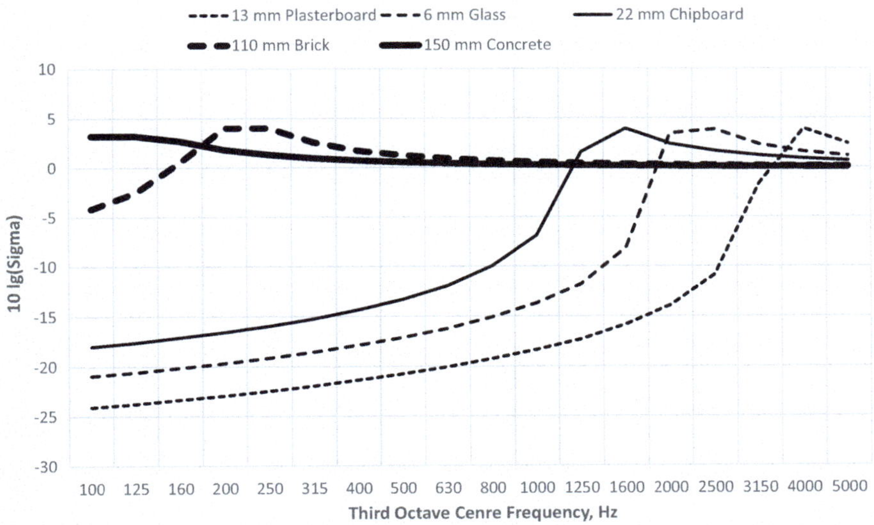

Fig. 7.2 Radiation efficiencies, expressed as 10 lg(σ), for common building materials

For light materials, 10 lg(σ) is low at low and mid frequencies, and above or close to 0 at high frequencies. For heavier materials, 10 lg(σ) is generally close to 0. This gives rise to the commonly used approximation for high-frequency noise radiated by a concrete slab:

$$\mathrm{SPL} \sim L_v - 27. \tag{7.5}$$

For direct excitation by a tapping machine, the high-frequency approximation for L_v can be combined with the formula for 10 lg(σ) to estimate the SPL resulting from the tapping machine operating on a bare surface, heard in a room directly below. Figure 7.3 shows the results for a bare 150 mm slab, compared with measured results from (Regupol, n.d.). The calculated level is within approximately 2 dB at 630 Hz and above, but below this frequency slab resonances and other effects come into play.

Exercise 7.1
By default, the "Plate SPL—Direct Excitation" sheet of the Supplementary file1 spreadsheet shows SPL values radiated into a standard room due to operation of a tapping machine on a bare 150 mm concrete slab. Repeat this for a bare 250 mm concrete slab. What is the difference in A-weighted levels produced?

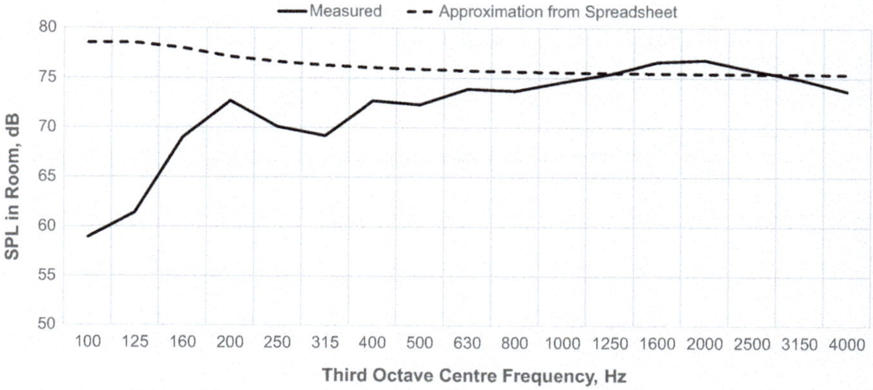

Fig. 7.3 SPL radiated due to a tapping machine on a 150 mm concrete slab—measured (data from (Regupol, n.d. ?)) and a calculated high-frequency approximation

7.6 Direct Assessment in Terms of SPL

In an existing building, a direct assessment of the significance of structure-borne noise can be obtained by simply measuring the SPL in an affected space due to a standard vibration source located in another space. For high-frequency vibration sources and horizontal plates, a standard tapping machine provides the source.

Procedures for the measurement are described in ISO 16283-2 (International Standards Organisation 2020a). In the basic procedure, the L_{eq} of the resulting sound is measured in octave or third-octave bands, and the resulting SPL is known as the *impact sound pressure level* L_i. A correction is usually added to account for the reverberation time in the receiving room, giving the standardised impact sound pressure level:

$$L'_{nT} = L_i - 10\,\lg(T/T_0), \tag{7.6}$$

where T is the reverberation time of the receiving room and T_0 is a "standard" reverberation time—for residential spaces $T_0 = 0.5$ s.

A single-number index can be formulated to summarise the measured values of L'_{nT} at different frequencies. This is defined in AS/NZS ISO 717.2 (International Standards Organisation 2020b) in a very similar way to the definition of $D_{nT,w}$ for airborne noise and is known as $L'_{nT,w}$. Correction terms can be defined for sources with different frequency characteristics, as for $D_{nT,w}$, but these are not commonly used because the vibration source for measurements is almost always a tapping machine which is largely a high-frequency source. A correction term C_I is defined in AS/NZS ISO 717.2 which accounts for the characteristics of certain types of floor, but its use is also rare. In practice $L'_{nT,w}$ is used almost exclusively to rate the impact isolation of floor/ceiling systems from high-frequency vibration sources such as footsteps.

A recent update to ISO 16283-2 introduces a 2.5 kg rubber ball, dropped from a height of 1 m, as a low-frequency source, with the resulting sound level measured as a maximum value with fast time-weighting rather than L_{eq}, and referred to as $L_{i,F}$. This could serve as a standardised test procedure for assessing low-frequency structure-borne noise, but at the time of writing, it is not widely used in Australia.

$L'_{nT,w}$ represents the sound level resulting from vibration of a floor when it is connected to other building elements in a constructed building. As for $D_{nT,w}$, a specification in terms of $L'_{nT,w}$ ensures that actual noise levels will achieve the desired goal and puts the onus on the builder to achieve the specification.

The index $L_{nT,w}$ is defined as for $L'_{nT,w}$ but is measured in a laboratory on a floor/ceiling assembly that is not connected to any other element. Hence, it can be used to rank-order different floor constructions, but does not necessarily represent the performance of the as-built construction. The difference between the laboratory-measured value and the value achieved in practice can often be very significant.

An alternative unit quoted by many manufacturers of isolation systems is $\Delta L_{n,r,0,w}$—often just written ΔL_w—which is defined in AS/NZS ISO 727.2 and roughly represents the difference in measured sound levels between a standard bare concrete slab and the same slab with the system under test. If the system is installed on a bare slab, $L_{nT,w}$ will be *approximately* $78 - \Delta L_w$. If not, the $L_{nT,w}$ of the complete floor/ceiling construction cannot be predicted. Once again, this metric is useful for rank-ordering the performance of isolation systems, but meeting a specification in terms of ΔL_w does not guarantee achievement of acceptable noise levels.

7.7 Control of Structure-Borne Noise

Most methods for control of structure-borne noise involve reducing vibration levels in the building structure, which is achieved by inserting one or more resilient elements in the chain between force input and SPL output.

The simplest such system is a mass on a spring (or other resilient element), with some damping, and this exhibits the basic properties of all isolation systems. Figure 7.4 shows the "force transmissibility" of this system—effectively the difference between force at the output and force at the input, in decibels. (Calculations can be found in the Supplementary file1 spreadsheet). The system has a resonant frequency f_0 at which there is usually amplification of the vibration signal. At higher frequencies, there is attenuation, which increases with increasing frequency.

The extent of both amplification and attenuation is determined by a "damping ratio" which reflects properties of the resilient material. Rubber, which forms the basis of many isolation systems, typically has a damping ratio of 0.1–0.2 depending on its composition.

The resonant frequency is given by

$$f_0 = \frac{1}{2\pi}\sqrt{\frac{\mathrm{k}}{\mathrm{m}}}, \tag{7.7}$$

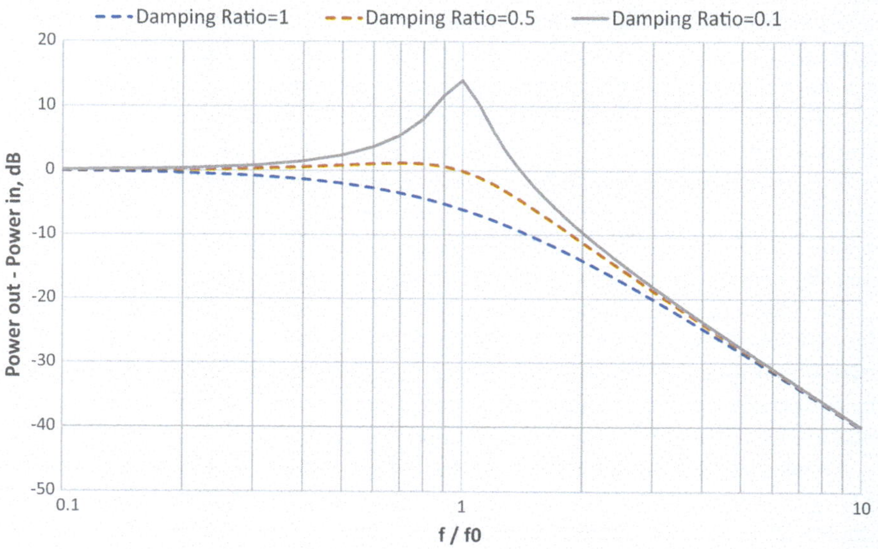

Fig. 7.4 Ratio of force at the output of a mass-spring system to force at the input (adapted from (Bies and Hansen 2009))

where k is the stiffness of the spring and m is the supported mass. Alternatively, this can be stated as

$$f_0 = \frac{0.5}{\sqrt{d}}, \tag{7.8}$$

where d is the deflection in *metres* of the spring with the load of the applied mass—for example, the amount by which rubber pads are compressed under the weight of a flooring system plus a person walking.

A number of principles for vibration isolation derive directly from this discussion.

- The resonant frequency of the vibration isolators must always be significantly below the frequencies present in the applied force. Hence, low-frequency sources such as dropping weights in a gym require isolators with higher deflection than high-frequency sources such as footsteps.
- Higher isolation performance (i.e., greater deflection) can be achieved either by using softer isolators or by loading them with higher mass. Very soft resilient elements may result in stability issues for the system. Increased load per isolator can be achieved for example by using point isolators rather than a mat. However, all isolation systems will have a maximum tolerable deflection that should not be exceeded.

- Damping in the system decreases the high-frequency performance but also decreases the amplification at resonance, providing some safeguard against "run-away" vibration levels if low-frequency vibrations were to occur. Standard rubber pads will have an amplification of about 10 dB at resonance.

Figure 7.5 shows a selection of vibration isolation systems designed for various purposes, including the typical deflection and, for high-frequency systems, the measured ΔL_w.

Exercise 7.2

(a) A fan has a blade-pass frequency of 50 Hz, and 20 dB reduction is required in structure-borne noise from the fan at that frequency. Rubber mounts will be used with damping ratio 0.15. From the "Mass-Spring Attenuation" sheet of the Supplementary file1 spreadsheet, find the required value of f/f_0. Hence, what should be the resonant frequency of the mounts?
(b) What is the required deflection of the mounts in mm?

7.8 Guidelines

For vibration other than footsteps and similar impacts on a floor directly above, the variation in types of vibration and mitigation measures means that simple assessment rules cannot be applied. If the space to be assessed does not yet exist, one needs to estimate the resulting vibration levels through the building, generally by measuring in similar structures, and use the principles outline above. From the vibration levels, the resulting SPL values can then be estimated. Assessment of the acceptability of that SPL can then be made as for airborne intruding noise.

For impact noise from a floor above, the Association of Australasian Acoustical Consultants has produced guidelines for appropriate values of $L_{nT,w}$ of the floor/ceiling system (Association of Australasian Acoustical Consultants, n.d.). In apartments and townhouses, the recommendations depend on the star rating of the building and are shown in Table 7.1. Other building types for which guidelines are provided are:

- commercial buildings,
- educational facilities,
- healthcare facilities.

In addition, the Australian National Construction Code recommends that in apartments, floor/ceiling systems between sole-occupancy units should achieve

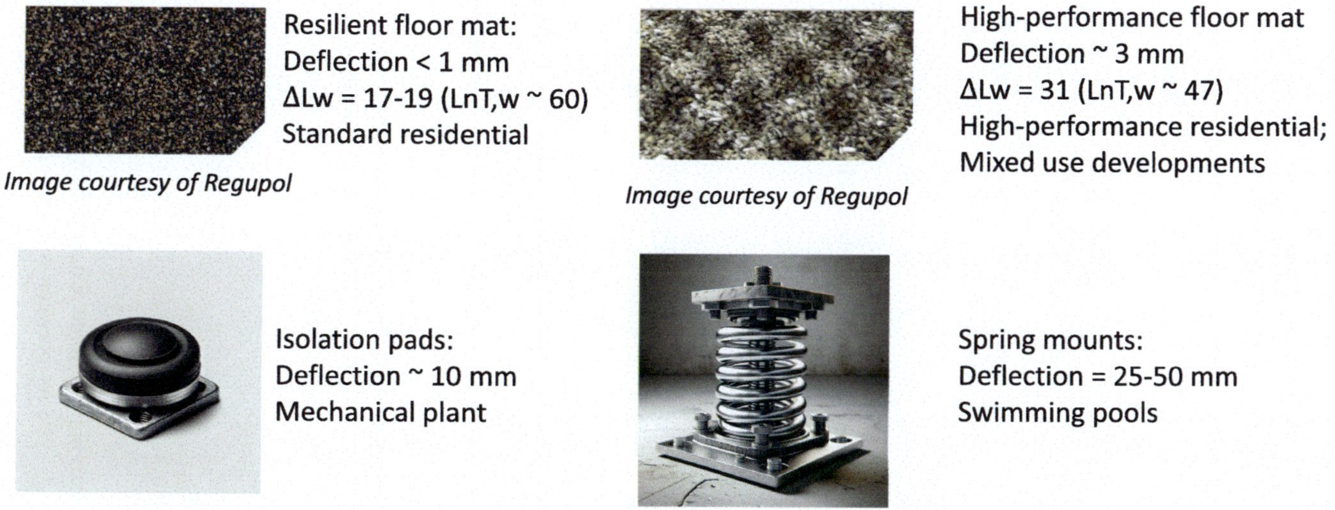

Fig. 7.5 Selection of vibration isolators

Table 7.1 AAAC recommendations for $L_{nT,w}$ values of floors between occupancies in apartments and townhouses (see (Association of Australasian Acoustical Consultants, n.d.))

Intertenancy activities	Star rating for sound intrusion				
	2 star	3 star	4 star	5 star	6 star
Between tenancies	65	55	50	45	40
Between all other spaces and tenancies	65	55	50	45	40

$L_{n,w} + C_1 \leq 62$ (approximately $L_{nT,w} \leq 62$). Comparison with Table 7.1 indicates that just meeting the NCC's conditions results in a space that the AAAC rates as between two and three stars.

7.9 Practical Considerations

As for airborne sound, there are many ways in which the performance of a vibration isolation system designed to control structure-borne noise can be compromised. These include:

- bridging elements between the isolated part of a system and the non-isolated part. For a flooring system, these may include hard connections between the floor and walls, or fastenings that connect the two elements.
- failure to take account of possible low-frequency resonant modes in certain building elements.
- failure to allow for possible lateral vibration of the isolated element, allowing it to come into contact with the non-isolated elements. This is usually handled by providing lateral restraints composed of very soft material.

As for airborne sound, site inspections are essential to ensure that works are being undertaken in accordance with recommendations.

Appendix 1: Propagation of Vibration in Buildings

Forms of Vibration

Vibration energy enters a structure through the application of a force. The force causes movement of the materials in the structure, and this movement propagates through the structure as waves.

There are several types of wave that can propagate in a structure. However, the most important in buildings is a *bending wave*. These waves can propagate through plates (like walls or floor slabs) and beams (like columns or studs). Figure 7.6 illustrates a bending wave in a plate.

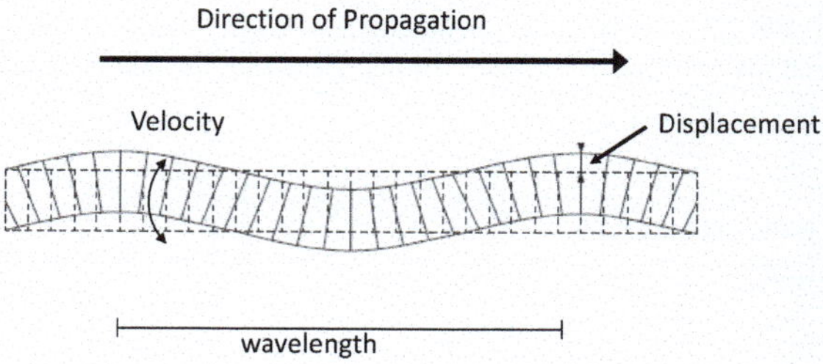

Fig. 7.6 Bending wave in a plate

In any form of vibration, particle movement can be described by:

- displacement (d),
- velocity (v),
- acceleration (a).

The time-average of the square of these quantities can be expressed in decibels:

$$L_d = 10\,\lg\left(\frac{d^2}{d_0^2}\right) \quad d_0 = 10^{-12}\text{m},$$

$$L_v = 10\,\lg\left(\frac{v^2}{v_0^2}\right) \quad v_0 = 10^{-9}\text{m/s},$$

$$L_a = 10\,\lg\left(\frac{a^2}{a_0^2}\right) \quad a_0 = 10^{-6}\text{m/s}^2.$$

Of these quantities, the most important is velocity, because it is directly linked to the energy carried by the wave:

$$E = m\langle v^2\rangle,$$

where $\langle v^2\rangle$ is the mean-square velocity and m is the mass of the plate.

Note that d, v and a are all vectors—they have components in the x-, y- and z-directions. The *resultant* is the vector sum of the three components—for example the resultant velocity is

$$v_r = \sqrt{v_x^2 + v_y^2 + v_z^2}.$$

For bending waves in a plate, the component in the direction normal to the plate is by far the most important.

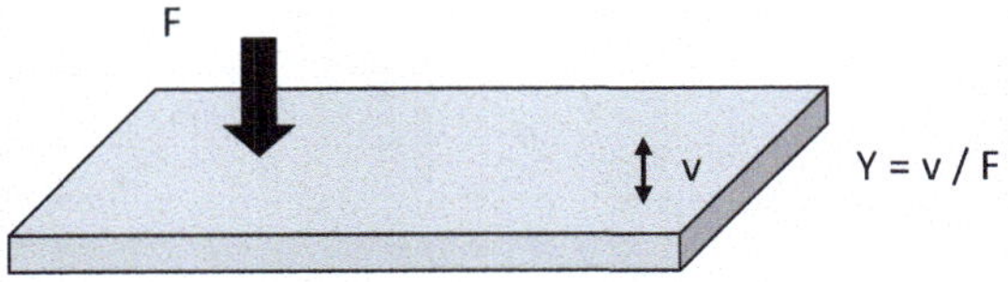

Fig. 7.7 Force, velocity and mobility

Velocity, Force and Mobility

The ratio of vibration velocity in a building element to the force required to produce it is the *mobility*, Y (Fig. 7.7). If the velocity is measured at the point where the force is applied, Y is called the *driving point mobility*. Otherwise, it is called the *transfer mobility*. Y is a complex number, but the important part is the real part Y_r.

For a finite-sized plate (wall, floor, etc.), mobility is in general quite complicated. It depends on how well the force "couples" to vibrational modes of the structure. At some locations and some frequencies, the structure resonates so that a force applied at that point with that frequency produces large velocities. At other points and/or frequencies, the mobility is lower.

This behaviour depends not only on the size and shape of the plate, but also on how it is connected to other parts of the structure. To calculate the mobility of a general structure requires a computer model, usually a "finite element" model.

Figure 7.8 from (Hopkins 2007) shows the driving point mobility of a 140 mm concrete slab. Curve (a) shows measured mobility averaged over the slab. At low frequencies, a number of resonances and anti-resonances are visible, which become less prominent at frequencies over a few hundred Hz. Curve (b) shows that a detailed model can give a reasonable prediction of the mobility. Curve (c) indicates the differences that can occur between different points on the same slab at specific frequencies.

For large plates and high frequencies, the driving point mobility approaches a value determined by the material properties:

$$Y_r \to \frac{1}{2.3\ \rho c_L h^2}, \tag{7.9}$$

where ρ is the density of the material, c_L is the speed of sound in the material and h is the thickness of the plate. This value is shown in Fig. 7.8, curve (a).

Internal Loss Factor

At points other than the driving point, vibration levels also depend on the amount of vibrational energy lost due to propagation through the plate. This is represented by the internal loss factor η. If the plate is not coupled to any others (so the only way that energy leaves is through internal losses), the total energy in the plate will be

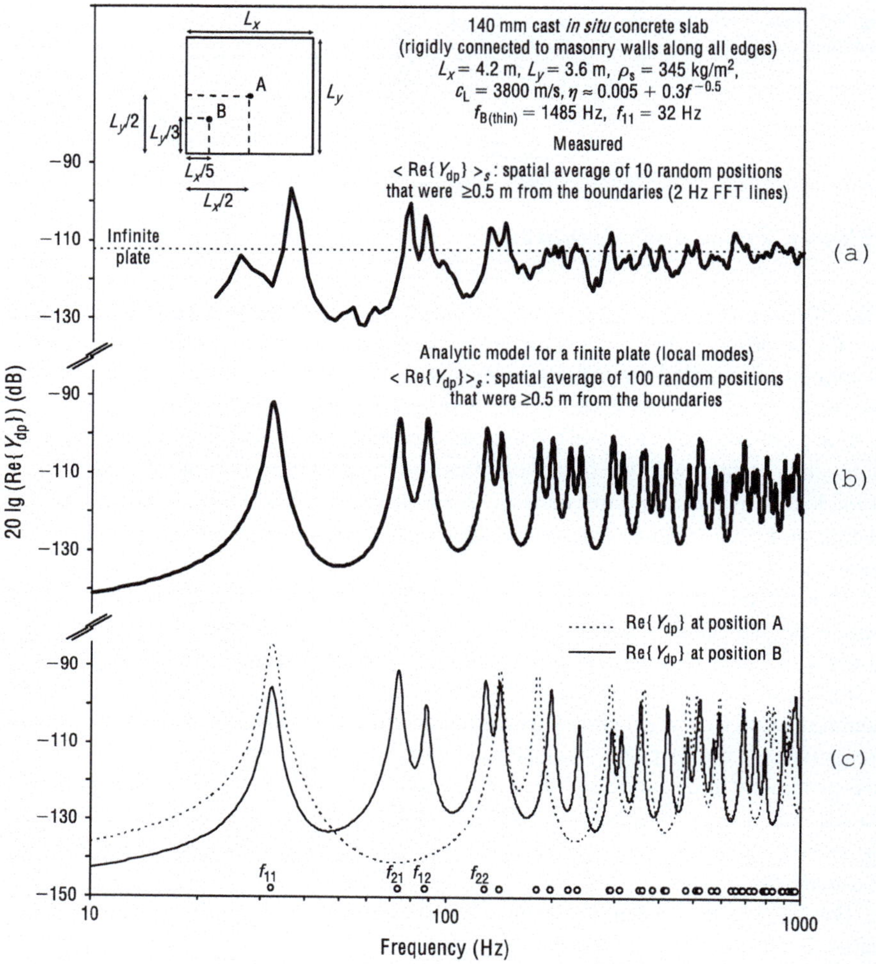

Fig. 7.8 Driving point mobility of a concrete slab: **a** average over slab, measured: **b** average over slab, model: **c** two specific points on slab (from Hopkins 2007)

$$E = m\langle v^2\rangle = F^2 \frac{Y_r}{2\pi f \eta}, \tag{7.10}$$

where F^2 is the mean-square force applied.

The internal loss factor is directly related to the material's damping. For thin plates, damping can be increased by introducing stiffening, which thereby reduces the vibration velocity for the same force input.

When the force applied is impulsive, the value of F^2 in any octave or third-octave band increases in proportion to the band centre frequency. For example, (Hopkins

2007) indicates that for a standard tapping machine, the value of F^2 in any third-octave band is $0.9f$, where f is the band centre frequency. Hence, the total energy in any third-octave band will be

$$E_{\frac{1}{3}\text{-octave}} = 0.9\frac{Y_r}{2\pi\eta}, \tag{7.11}$$

which is independent of frequency provided that Y_r is constant.

Transmission of Vibration to Other Building Elements

When energy is transferred from one vibrating element to another, it is also subject to losses. These are called *coupling loss factors*. Once again these are very difficult to predict in a specific building without a finite element model. They are typically between zero and 10 dB for transmission between two elements. However, a conservative rule of thumb that generally applies at high frequencies is:

- Where a number of building elements are in contact with a vibrating element, assume the vibration energy—that is, $m\langle v^2\rangle$—is equally distributed between elements. Hence, elements with lower mass will have higher vibration velocities.

Vibration energy can also be transmitted into the building structure from the surrounding ground (e.g. train vibration). The coupling loss between the ground and the structure can typically be taken to be:

- about 5 dB for a structure on piles,
- 0 dB for a slab on the ground.

References

Association of Australasian acoustical consultants. guidelines and downloads, [Online]. Available: https://aaac.org.au/Guidelines-&-Downloads

Bies D, Hansen C (2009) Enginering noise control, 4th edn. Spon Press

Hopkins C (2007) Sound insulation. Elsevier

International Standards Organisation (2020a) ISO 16283-2:2020 acoustics—field measurement of sound insulation in buildings and building elements. Part 2—impact sound insulation. International Standards Organisation

International Standards Organisation (2020b) ISO 717-2:2020 acoustics—rating of sound insulation in buildings and of building elements. Part 2—impact sound insulation. International Standards Organisation

Regupol. Download test reports, [Online]. Available: https://www.abswest.com.au/index.php/products/flooring-and-underlay/acoustic-underlay/regupol-sonus-core-5-acoustic-underlay/#test-reports

Chapter 8
Sound Quality

Abstract This chapter describes how to assess the quality of sound in a space, how to determine whether that is consistent with the purpose of the space, and how to specify acoustic parameters to achieve a required sound quality. Much of the discussion concerns reverberation—its subjective impact, measurement and design goals. Other descriptors are also discussed, including clarity, coloration and the effect of echoes. The chapter concludes with a summary of good design practice for specifying sound quality.

8.1 Sound Level Versus Sound Quality

The sound pressure level of sound in a room, either overall or as a function of frequency, basically describes the loudness of the sound. This is important for assessing the appropriateness of background sound and the significance of intruding noise in relation to the background. However, there are many other aspects of our experience of sound in spaces. These are described in more aesthetic terms such as

- reverberation— "live" versus "dead",
- clarity— "crisp" versus "muddy",
- echoes—present or not,
- coloration— "warm" versus "cold",
- spatial distribution— "enveloping" versus "hollow",
- other un-nameable properties that give a space (or a position in a space) a particular "personality".

Most of the above descriptions can be more or less captured in numerical terms by considering properties of the *impulse response* at a listening point, or in some cases two impulse responses, one at each of the listener's ears. The impulse response of a space is described in Chap. 2, Appendix 2.

Supplementary Information The online version contains supplementary material available at https://doi.org/10.1007/978-981-97-9380-8_8.

R. Bullen, *The Architecture of Sound*, Springer-AAS Acoustics Series,
https://doi.org/10.1007/978-981-97-9380-8_8

In this Chapter, the above listed features are considered, including criteria and guidelines for various types of space and design principles. Computer simulation of sound quality is described, and finally, Sect. 8.8 summarises what would currently be considered good practice in designing for sound quality.

The absorption and diffusion properties of building materials are described in Appendix 1 to this chapter and will be referred to in this section.

Note that one other very important aspect of sound quality—speech intelligibility—is discussed separately in Chap. 9.

8.2 Reverberation

8.2.1 Objective Indicators of Reverberation

Reverberation is the most significant of the above factors in determining the sound quality in any space. It is quantified by the reverberation time (RT), which as described in Chap. 2, Appendix 2 represents roughly the time for the sound level produced by a very short "impulse" of sound to decay by 60 dB. More precisely, it is calculated from the slope of the decay curve of SPL vs. time and has a number of variants:

- reverberation time as defined in AS 2460:2002, which is based on the slope of a decay curve in the section where the slope is linear. This generally represents "late" decay and is referred to here simply as *RT* ,
- RT_{20}, based on the slope of the decay when the SPL is between −5 dB and
- −25 dB relative to the maximum level, but extended linearly to give the time to decay by 60 dB,
- RT_{30}, similarly defined between −5 dB and −35 dB,
- Early Decay Time (*EDT*) based on the slope between the maximum level and the maximum level minus 10 dB.

RT_{20} and RT_{30} are defined in International Standard ISO 3382.

All the above versions of RT depend on frequency and are usually quoted at octave-band centre frequencies. When the generic term "reverberation time" is used without specifying the frequency, this usually means RT_{20} or RT_{30} at mid frequencies—an average of values in the 500 Hz and 1 kHz octave bands.

8.2.2 Subjective Impact of Reverberation

We have already met RT in Chap. 2, in the context of determining the SPL in a Sabine space. Doubling the RT in such a space will increase the reverberant-field contribution to SPL by 3 dB for all sounds. However, the effect of reverberation on sound quality is much more significant than its effect on sound level. Reverberation changes the "feel" of a space, particularly the sense of size and intimacy as well as the sense of peacefulness.

An art gallery with an RT of 1.3s will be perceived as formal and imposing, suited for large works viewed from a distance, as in this simulation {Link to Supplementary file2}.

An art gallery with an RT of 0.5s will be perceived as intimate, encouraging close inspection of smaller works, potentially over a long period, as in this simulation [Link to Supplementary file3} which uses the same sounds.

A café with an RT of 1.1s [Link to Supplementary file4} will be "buzzy" and lively. People will speak loudly and will not stay for long periods.

A café with an RT of 0.3s [Link to Supplementary file5} will feel more restful, encouraging quiet speaking and longer stays.

For music performance, there are specific requirements. Much "classical" music was written to take advantage of the acoustic properties of the spaces where it would be played, which typically have reverberation times of 1.5–2s, or even higher for choral and organ music. Recommendations for RT in spaces for unamplified music performance are provided below.

On the other hand, amplified music is generally rhythm-dependent and needs a clear perception of the beat, which can become confused in a highly reverberant environment. Reverberation can be added by the sound engineer if required, so the ideal environment for this type of music is a space with a low reverberation time.

For clear speech listening, relatively low reverberation times are required. In a large auditorium, it is often recommended that RT is about 0.8–1.2s to increase the SPL of the voice. However, if background SPLs can be low enough to prevent masking of the wanted speech, keeping the RT as low as 0.5s increases the "presence" and directness of the voice. Compare the sound of speech in the lecture room [Link to Supplementary file6} from Chap. 2, Appendix 2 (RT = 1.2s @ 1 kHz) to that in the Hotel Room [Link to Supplementary file7} (RT = 0.6s @ 1 kHz).

Design of spaces for speech intelligibility is described in detail in Chap. 9.

8.2.3 Reverberation Criteria

Appropriate reverberation times for general spaces can be based on recommendations in AS/NZS 2107:2016 (Standards Australia 2016). As for background sound levels, this standard gives a range of RT values that would generally be considered appropriate for each space. Table 8.1 shows examples of recommended values for office spaces. These recommendations are valuable but must be considered in the light of the specific design intention for each space. The two café environments in the examples given above, with very different reverberation times, may each be appropriate depending on the ambience sought. In addition, Chap. 9 describes cases where the recommendations in Table 8.1 may not be compatible with good speech intelligibility.

Table 8.1 Recommendations from AS/NZS 2107 for reverberation times in "Office Buildings"

Type of occupancy/activity	Design reverberation time range, s
Board and conference rooms	0.6 to 0.8
Cafeterias	< 1.0
Call centres	0.1 to 0.4
Corridors and lobbies	< 1.0
Executive office	0.4 to 0.6
General office areas	0.4 to 0.6
Meeting room (small)	< 0.6
Open plan office	0.4*
Public spaces	0.5 to 1.0
Quiet rooms	< 0.6
Reception areas	0.6 to 0.8
Rest rooms and break-out spaces	0.4 to 0.6
Video/audio conference rooms	To 0.4

*Reverberation time should be minimised for noise control

For spaces where acoustics is more critical, AS/NZS 2107 suggests mid-frequency reverberation times which depend on the room volume, as shown in Fig. 8.1. It notes that these are intended as guides only, as spaces considered to have good acoustic qualities may have significantly different values of RT. Note also that in this context "music" means "unamplified music".

Many authors have suggested alternative or additional recommendations for reverberation time, particularly in concert halls. However, it is now recognised that in

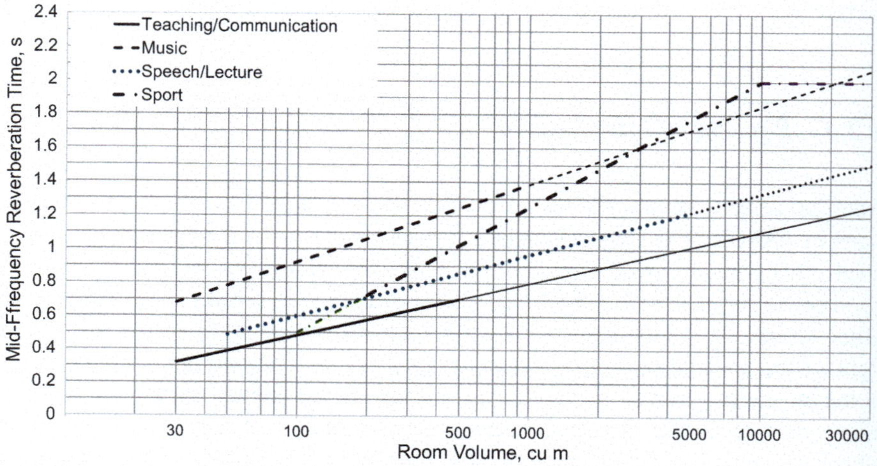

Fig. 8.1 Suggested mid-frequency reverberation times, adapted from AS/NZS 2107:2016

spaces where acoustics is critical, the other factors listed in Sect. 8.1 may be just as important as RT in achieving the desired acoustic environment.

8.2.4 Calculating Reverberation Time

As explained in Chap. 2, Appendix 2, not every space can be said to have a single reverberation time that applies everywhere—this applies only to "Sabine" spaces where absorption is low enough and/or reflections are sufficiently randomised that a diffuse sound field is formed. They include auditoria and other spaces where reverberation is significant. (Chap. 2, Appendix 2 gives a more precise definition of a Sabine space.) Where these conditions are met, RT can be estimated by the Sabine reverberation time:

$$RT_S = .161\frac{V}{S\alpha}, \tag{8.1}$$

in which S and V are the total surface area and volume of the space in SI units, and α is the mean absorption coefficient of the space—that is:

$$\alpha = \frac{\sum_i S_i\alpha_i}{S}, \tag{8.2}$$

where S_i and α_i are respectively the surface area and absorption coefficient of each surface in the space. (Absorption properties of surfaces are described in detail in Appendix 1 to this chapter). The units of $S_i\alpha_i$ are m^2, but in this context, they are often called Sabines or m^2 Sabines.

In spaces with slightly higher mean absorption (α greater than about 0.05) but still complying with the other requirements for a Sabine space, reverberation can be described by the Norris-Eyring reverberation time:

$$RT_{N-E} = .161\frac{V}{S(-\ln(1-\alpha))}. \tag{8.3}$$

This equation has the advantage that as α approaches 1, it correctly predicts a value of 0 for the reverberation time. However, both Eqs. (8.1) and (8.3) are in fact based on assumptions which include α being small.

Two refinements of Eqs. (8.1) and (8.3) are commonly used. First, room contents such as unoccupied chairs or people do not have an easily defined surface area. These are generally assigned an "absorption area" for each item in Sabines, to be included in the sum in the numerator of (8.2). (Sometimes a large area of seating is assigned an equivalent absorption coefficient which can be multiplied by the seating area as for a surface.) Second, in large spaces air absorption can add significantly to the total absorption at high frequencies. Taking these factors into account, Eq. (8.2) becomes

$$\alpha = \frac{A_T}{S}, \tag{8.4}$$

where A_T is the total absorption, given by

$$A_T = \sum_i S_i\alpha_i + \sum_j A_j + 0.92m'V. \tag{8.5}$$

Here

- A_j is the absorption area of each of the room contents, i.e. chairs, etc.
- m' is air absorption in dB/m calculated according to ISO 9613-1.

As an example of the significance of air absorption, in a room with a volume of 200 m^3, the third term in Eq. (8.5) will add an absorption area of approximately 1 m^2 at 1 kHz or 6 m^2 at 5 kHz.

It is notable that Eq. (8.1) does not depend on the location of the absorption or the shape of the space. In a Sabine space, the reverberation time is simply a function of the volume of the space and the total amount of absorption in it.

Of course, most real spaces are not Sabine spaces. Nevertheless, entering the total absorption in any space into Eq. (8.1) generally gives a value that is representative of the level of perceived reverberation as experienced by a listener, and can be compared with criteria such as those in Sect. 8.2.3.

Another descriptor of reverberation is the Fitzroy reverberation time. This can be seen as an alternative way to estimate the perceived reverberation experienced in non-Sabine spaces where the departure is due to non-random distribution of absorption in the space. The space is considered to consist of three sets of two parallel planes, representing the x-, y- and z- axes, having areas S_x, S_y and S_z and average absorption coefficients α_x, α_y and α_z. The "reverberation" is described by

$$RT_F = .161\frac{V}{S^2}\left\{\frac{-S_x}{\ln(1-\alpha_x)} + \frac{-S_y}{\ln(1-\alpha_y)} + \frac{-S_z}{\ln(1-\alpha_z)}\right\}. \tag{8.6}$$

If $\alpha_x = \alpha_y = \alpha_z$, then $RT_F = RT_{N-E}$. Otherwise RT_F is always larger.

A less commonly used descriptor known as the Arau-Puchades reverberation time also accounts for non-random distribution of absorption (Arau-Puchades 1988).

These estimates of reverberation can form the basis of acoustic design for acoustically non-critical spaces, with other "sound quality" factors discussed in Sects. 8.3 to 8.6 being considered in cases where a higher standard of sound quality is desired. Most commonly, in non-critical spaces perceived reverberation is estimated using Eq. (8.1).

In critical listening spaces, the other "sound quality" factors are of greater importance. However before discussing these, two points should be made regarding the use of Eqs. (8.1), (8.3) or (8.6).

What Counts as Absorption?

- If absorbing material is placed against another surface, the absorption coefficient of the material multiplied by its area is added into Eq. (8.4), but generally not the absorption of the surface that it covers. (Exceptions occur only for very light covering materials).

- If both sides of a solid object are exposed to sound in a room, e.g. a hanging or partial-height partition, then absorption can be counted from both sides, so the total absorption is the twice the object's area times the absorption coefficient of the material.

When is a Room a Room?

In the very common case where two spaces have an opening between them, there are two extremes:

- If the opening is small enough, the two spaces can have different reverberation times and different perceived reverberation environments. The opening will count as absorption in both spaces.
- If the opening is large enough, the two spaces are acoustically one space with reverberation characteristics set by the total volume and total absorption area in both spaces.

In intermediate cases, the reverberation curve will not be linear, and full analysis requires tools described in (Cremer and Muller 1978). However, application of those tools to typical room sizes and absorptions suggests the following "rule of thumb":

- The reverberation characteristics of Space 1 will not be significantly affected by those of another coupled space (Space 2) provided that the area of the opening between them is less than 1–2% of the total surface area of Space 1.

Otherwise, it is best to consider the two spaces as one, but be aware that there will be some difference in the quality of sound in each space if the amount of absorption in the two areas is very different.

For example, a classroom with dimensions $10 \times 8 \times 2.8$ m would have a total surface area of 260.8 m^2. A storeroom with a door open to the classroom could affect its reverberation if the opening has dimensions greater than about 2.6 m^2.

Exercise 8.1

A "shoe-box" shaped room has dimensions 5 m $\times$ 10 m $\times$ 3 m. The surfaces and their absorption coefficients at 500 Hz are:

Surface	Material	α at 500 Hz
Ceiling	Standard tiles fixed to soffit	0.5
Floor	Medium carpet	0.3
Side walls (10 m $\times$ 3 m)	"Acoustic" plaster	0.2
End walls (5 m $\times$ 3 m)	Timber panels	0.05

To the nearest 0.1 s, at 500 Hz, what is the calculated:

(a) Sabine reverberation time
(b) Norris-Eyring reverberation time
(c) Fitzroy reverberation time

8.3 More Complex Descriptors of Reverberation

8.3.1 Differences Between Reverberation Time Indicators

When the decay curve of SPL versus time is not linear, the four "reverberation time" indicators listed in Sect. 8.2.1 will be different, sometimes substantially so. The first variant, referred to here simply as *RT*, will estimate the slope of the reverberation curve towards the late stages, where the sound is "jangled". The other three values give various indicators of the earlier stages of the decay, where the reverberation signal contains more information from the source.

Reasons for curved sound decays include:

- non-random location of absorption, often concentrated in the ceiling,
- coupled or semi-coupled spaces with different levels of absorption, as described in "When is a Room a Room?" above,
- at low frequencies, room modes as described in Sect. 8.5,
- a source or receiver location close to a reflecting surface.

When a decay is nonlinear, it will also vary with location in the space, so there is not one value of any of the reverberation descriptors that applies to the whole space.

Figure 8.2 compares the "Hotel Room" and "Shared Office" environments described in Chap. 2, Appendix 2. For the "Shared Office", the reverberation curve is almost linear, while for the "Hotel Room", it is significantly nonlinear.

Table 8.2 indicates that the Hotel Room has a higher *RT* value, representing the slope in the later stages of the decay, but lower values of other reverberation indicators, representing earlier stages of the decay.

Samples of speech as simulated in each of these environments can be heard by clicking in Table 8.2. In the "Hotel Room" environment, the speaker sounds closer (although both are recorded at about 2 m) and there is less subjective reverberation.

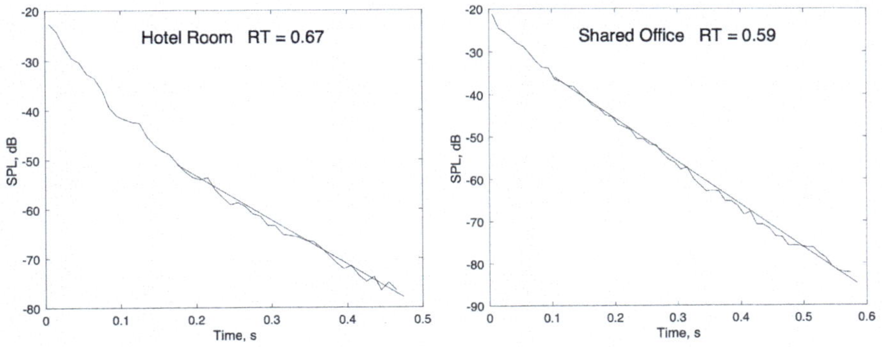

Fig. 8.2 Reverberation decays for the "Hotel room" and "Shared office" environments described in Chap. 2, Appendix 2

Table 8.2 Reverberation time parameters for two example spaces, secs (Open the associated Supplementary file to hear speech sample)

Example space	EDT, s	RT_{20}, s	RT_{30} , s	RT , s
Hotel room [Link to Supplementary file8}	0.27	0.37	0.42	0.67
Shared office [Link to Supplementary file9}	0.35	0.51	0.53	0.59

8.3.2 Clarity and Gain

Two other descriptors are often more useful than reverberation time in describing the "feel" of a non-Sabine space. These are described in Chap. 2, Appendix 2, and are defined in International Standard ISO 3382-1, Annex A. They are:

- Clarity, or C_{50} , representing the ratio of "early-arriving" to "late-arriving" energy in a decay. Higher values indicate greater "presence" and intimacy of the sound. In a Sabine space, clarity is simply a function of reverberation time and distance from the source, but in other spaces it offers information in addition to both early and late indicators of reverberation.
- Gain is the difference between the SPL of a constant sound at the receiver location and the level of the same source at 10 m in a free-field environment. Gain is an important parameter used in the design of large auditoria for unamplified music and in this context is often called Strength. Higher reverberation times generally produce higher values of gain, but lower values of clarity.

Table 8.3 compares two spaces which both have large values of RT_{30} but very different values of clarity. (Gain cannot be computed from the example impulse responses because they are not scaled—see Chap. 2, Appendix 2). The difference in sound quality is clear from the speech samples.

Table 8.3 RT30 and clarity for two example spaces (Open the associated Supplementary file to hear speech sample)

Example space	RT_{30}, s	Clarity, dB
Concert hall [Link to Supplementary file10}	1.89	0.4
Stairwell [Link to Supplementary file11}	1.48	−7.7

8.4 Echoes

An echo is a sudden increase in sound from the source, occurring more than about 50 ms after the arrival of the original sound. Certain room shapes, particularly those involving curved surfaces such as in Fig. 8.3, can result in focussing of sound in one location, resulting in uneven distribution of sound throughout the space and in some cases an echo at certain locations. A detailed discussion can be found in Vercammen (2010a) and Vercammen (2010b). Another example is a "flutter echo" due to multiple reflections from parallel hard surfaces.

Echoes should not be confused with reverberation—reverberation can be good or bad depending on the purpose of the space, but echoes are almost always bad for the listening environment.

Here is an example of speech in the "Hotel Room" environment with an echo [Link to Supplementary file12} occurring at 100 ms after the original sound.

In the absence of a computer model, avoidance of echoes is best achieved by ensuring that:

- any surfaces that are curved as in Fig. 8.3 are absorptive,
- any non-absorptive surfaces, particularly those facing other hard surfaces, incorporate significant diffusion (described in Appendix 1 to this chapter).

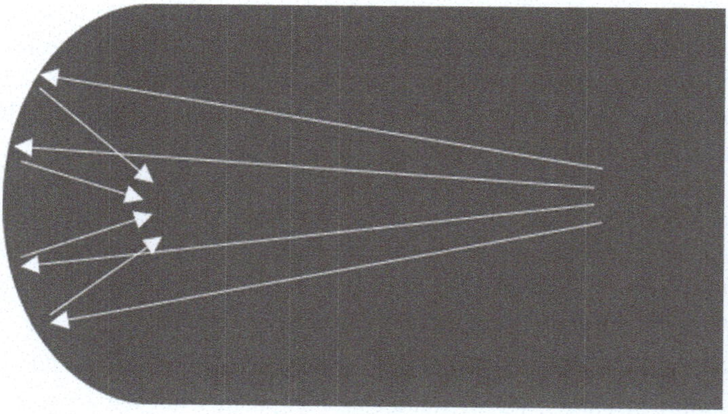

Fig. 8.3 Example of a geometry that could lead to an echo

8.5 Colouration

Because reverberation depends on frequency, sound heard in any space will have some frequencies emphasised and some de-emphasised. The effect is known as "coloration" and can be seen in the frequency dependence of measured reverberation times (see Chap. 2, Appendix 2). A more direct way is to take the Fourier transform of the impulse response, which gives the space's *frequency response* (or rather the frequency response at one specific receiver position). This gives a much finer indication of frequencies that may be emphasised or de-emphasised.

Of course, the degree to which this measured frequency response is representative of the entire room depends on the representativeness of the impulse response, which depends on the location of the source and receiver as well as the directional characteristics of the source at each frequency.

Coloration is generally due to one of three factors.

- At low frequencies, the shape and size of the room can generate resonances at specific frequencies, known as "room modes". This is particularly the case for small rooms. Prediction and control of room modes is described in many texts, e.g. (Hopkins 2012).
- The absorption properties of room surfaces and contents can result in higher absorption in some frequency bands.
- Multiple non-absorptive surfaces with no diffusion can cause interference effects which produce high-frequency variations in the room response.

Figure 8.4 shows examples of frequency responses calculated from recorded impulse responses in the "Hotel Room" and "Concert Hall" environments described in Chap. 2, Appendix 2. Clearly the Hotel Room exhibits modal resonances at low frequencies and more variation at mid-to-high frequencies.

Here is an example of music as heard in the Hotel Room [Link to Supplementary file13} and the concert hall [Link to Supplementary file14} (using a

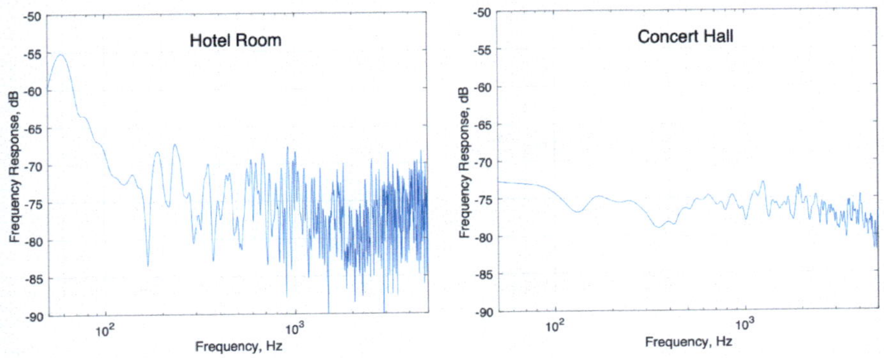

Fig. 8.4 Frequency responses calculated from impulse responses in the "Hotel room" and "Concert hall" environments

music sample sourced from http://www.lam.jussieu.fr/Projets/index.php?page=AVAD-VR). Sound in the concert hall has longer reverberation and much less coloration. In the "Hotel Room" environment, the sound is "dead" and "muddy". (Note that these are simple simulations using a single impulse response—a full computer model of the space would include the effect of the listener's head and other effects.)

These considerations are relevant in spaces where listening to music (amplified or unamplified) is the prime purpose of the space. However, as described in Chap. 9, they are also important when speech intelligibility is critical.

8.6 Spatial Distribution

Where listening is the major purpose of the space, the range of directions from which sound reaches a listener is also significant, although the importance of this factor, objective indices and criteria are all subject to much debate. Some general rules that have been proposed include the following:

- For (unamplified) music listening, sound should arrive primarily from the side of the listener rather than from above ("early lateral reflections"). This gives a sense that the listener is enveloped in the sound.
- For speech listening, any reflected sound should arrive from above, which promotes the impression that it arrives directly from the speaker. These two rules lead to the following general advice: for music, locate absorption on the ceiling; for speech, locate it on the walls (provided this does not cause echoes or coloration).
- "Inter-aural cross correlation" represents the similarity between the sound signals reaching the listener's two ears. When this is low, there is a sense of spaciousness—the source seems to cover a large area. This is considered by some practitioners to be good, especially for music listening.

Once again, these considerations are relevant only for very high-end applications, and in those cases, detailed design requires more understanding than can be provided in this book.

8.7 Computer Simulation Programmes

In circumstances where modes are significant, as in the "Hotel Room" in Fig. 8.4 at low frequencies, the only way to accurately model the acoustic performance of a room is to use a computer-based numerical technique such as "finite element analysis" (FEA) or "finite difference time domain" (FDTD). These are described in standard texts including (Ochmann and Mechel 2004). These techniques can identify the likely resonant frequencies of any room shape, and to some extent can also identify the level of amplification to be expected. However, they are very computing-intensive and at high frequencies, the results are extremely sensitive to assumptions about the room surfaces and contents. In practice these methods are often used to calculate the impulse response of a room at low frequencies, although their use is largely restricted to high-level applications such as recording and mixing studios.

At higher frequencies, the most accurate way to calculate the impulse response is to calculate images of the source in each room surface, then images of each image source, etc., and then trace the reflections resulting from each image source, including any "smearing" due to absorption and diffusion effects.

In practice this is currently too computation-intensive for most applications, and most modelling programmes use "ray tracing"—following the paths of a large number of sound "rays" emitted from the source and noting when they pass close to the receiver. This is an approximation to image sources and has a number of theoretical problems, but in practice it appears to work quite well.

Programmes using ray tracing include EASE and Odeon, although Odeon combines the image source and ray tracing methods. They produce a direct calculation of the point-to-point impulse response, or sometimes two impulse responses, including the effect of the listener's head. These can then be used to produce RTs and other measures, as well as audible simulations using recordings made with no reverberation. Such programmes are often intended for use in sound system design, although they can be used for acoustic design more broadly. Details of the techniques involved in producing audible simulations can be found in Vorlander (2008).

8.8 Summary of Good Design Practice

At the design stage of a project, current good practice in terms of design for sound quality could be summarised as follows:

- For non-critical spaces: select a value of RT using AS/NZS 2107 as a starting point, but tailor this according to the design intent for the space. That is, consider whether the space should be peaceful or "buzzy", whether conversation should be easy to understand, etc., and increase or decrease the selected value accordingly. Having determined an appropriate RT, design to achieve it using the principles in Sect. 8.2.4.
- For spaces where listening to sound is the main function of the space—generally music venues and speech/music auditoria, but also including high-end conference rooms and similar: select a value of RT as above, and also select criteria for other indicators of sound quality in Sects. 8.3–8.6. Prepare a first-pass design using the basic principles in those sections. Then submit this to computer modelling, with the model including sounds intended to be heard as well as any background sound. Check that the numerical criteria are met in the model, and listen to an audible simulation of sound in the room. Adjust the design until you (or your client) are satisfied that the sound is appropriate.

Appendix 1: Sound Absorption and Diffusion by Materials

The Sound Absorption Coefficient

All materials have an absorption coefficient, α, between 0 and 1, which is the ratio of the sound power absorbed by the material (that is, not reflected back into the space from which it came) to the sound power incident on it. The absorbed power is converted to other forms of energy, as described below. (For the purpose of building acoustics, power transmitted through the material is also counted as being absorbed. However, in most practical cases, this is a tiny fraction of the total incident power).

An absorption coefficient of 1 means that the material will absorb 100% of the incident sound power; an example of such a "material" is the open area of a window for which all incident power is transmitted through. Conversely, an absorption coefficient of 0 means that the material will reflect 100% of the incident sound power.

Like the sound reduction index R described in Sect. 4.2 of the main text, α depends on the frequency of the sound and also on its angle of incidence. In general, α is higher at angles close to normal incidence. Also like R, we can speak of a "normal incidence" and a "random incidence" value of α, where "random incidence" implies an average over all incidence angles.

The absorption coefficient of a material can be measured in two ways.

a. In an "impedance tube", sound is directed at the material and the ratio of incident to reflected power is determined from the pattern of sound pressure generated in the tube. This gives the normal incidence value of α.
b. In the reverberant method, a sample of material is placed in a room designed to be a Sabine space. Assuming that the area of the material is small and the space remains Sabine when the absorbing material is added, the random-incidence value of α will be given by:

$$\alpha_{\text{rand}} = \frac{.161V}{S}\left[\frac{1}{T} - \frac{1}{T'}\right] \tag{8.7}$$

where

- S is the area of the sample.
- V is the room volume.
- T is the reverberation time with the sample present.
- T' is the reverberation time in the absence of the sample.

  A slightly more complex equation is given in ISO 354:2003 (International Standards Organisation 2003) which allows for the possibility that atmospheric conditions in the test room may differ between the two measurements.

This is the form of α that is most commonly reported, and when an "absorption coefficient" is referred to this will invariably mean α_{rand}.

Unfortunately, for highly absorbing materials, the assumption that a test space remains Sabine with the material present breaks down, and hence, the resulting value of α is not accurately described as "random incidence".

In addition, diffraction effects occurring at the edges of the sample mean that the effective absorbing area may be greater that the measured area. This can lead to a number of inaccuracies, including calculated values of α_{rand} greater than 1 when using Eq. (8.7) (Sauro and Mange 2009). These values are often reported in manufacturers' literature, but in most spaces using values of α_{rand} greater than 1 will over-estimate the total absorption in the space. In calculations, many practitioners replace reported values of α_{rand} that are greater than 1 with 1.

The "Noise Reduction Coefficient" or NRC of a material is simply the average value of α_{rand} for octave bands from 250 Hz to 2 kHz. (The unit is poorly named—it represents absorption, not noise reduction).

Porous (or Dissipative) Sound Absorbers

In these materials, sound energy enters small passages in the material and is converted to tiny amounts of heat. Commercially available porous sound absorbing materials include rockwool, fibreglass and polyester. Porous absorbers also include common materials such as curtains or blinds, carpets, soft furnishings and people (or rather their clothing).

A list of measured absorption coefficients can be found in (Bies and Hansen 2009, Sect. 7.7.1). Figure 8.5 shows a selection of measured values of α_{rand} for porous absorbers.

It is clear from Fig. 8.5 that porous absorbers operate largely at high frequencies—the only exception is ceiling tiles suspended below the ceiling, which is a system that also functions as a panel absorber at low frequencies (see below). Factors affecting the performance of porous absorbers include the following.

Thickness

The effect of thickness on the absorption coefficient of fibreglass or rockwool is illustrated in Fig. 8.6. A similar relationship applies to all porous absorbers.

Airspace

Low-frequency absorption can be improved by including an air gap between the material and the reflective wall behind.

Wrapping or Bagging

To protect a porous material from contamination, or to control potential health impacts from inhalation of fibres, it is possible to cover it with a thin, limp, impervious plastic film. When the insulation is wrapped or bagged, this generally increases its low-frequency absorption slightly but decreases its high-frequency absorption. If the wrapping material is very thin (such as 20 μm polyurethane), then the effect on

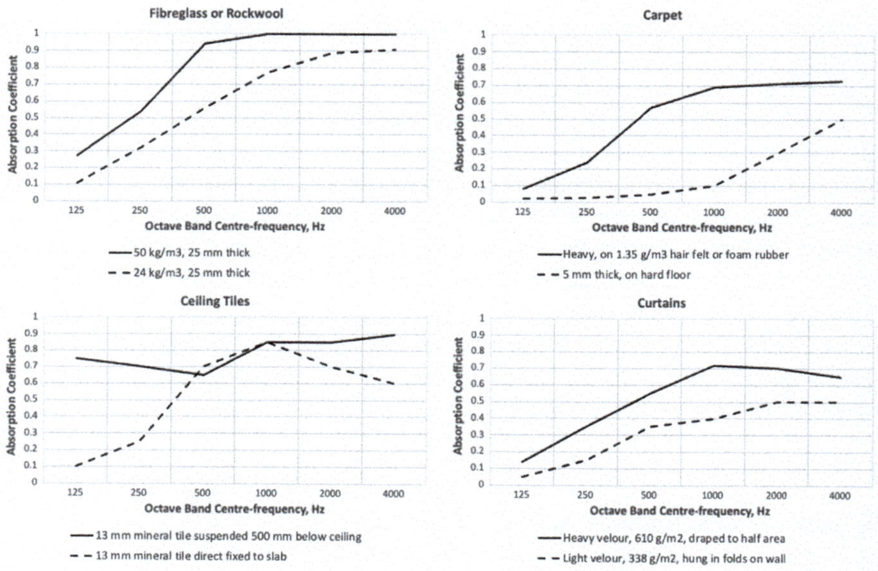

Fig. 8.5 Measured values of α_{rand} for a selection of porous sound absorbers (from Bies and Hansen 2009)

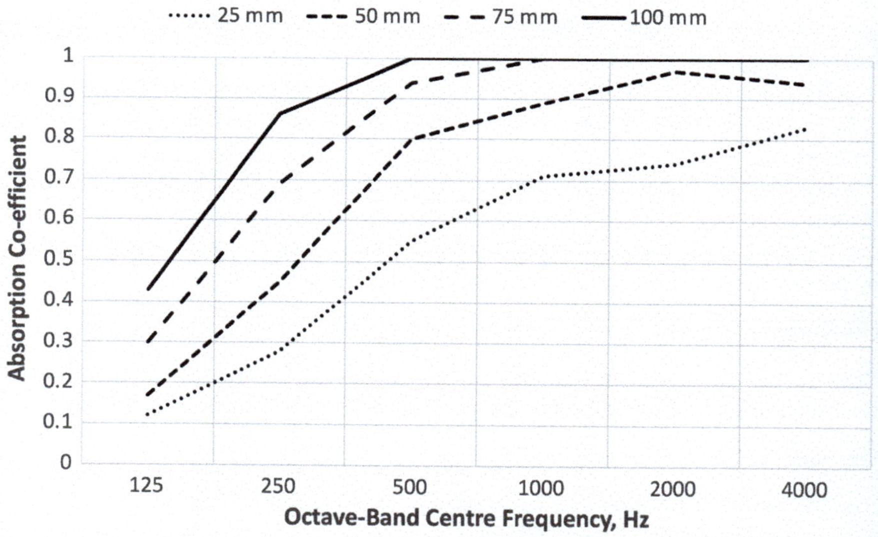

Fig. 8.6 Measured values of α_{rand} for 16 kg/m^3 fibreglass or rockwool with the thicknesses shown (from Bies and Hansen 2009)—measured values truncated at 1.0)

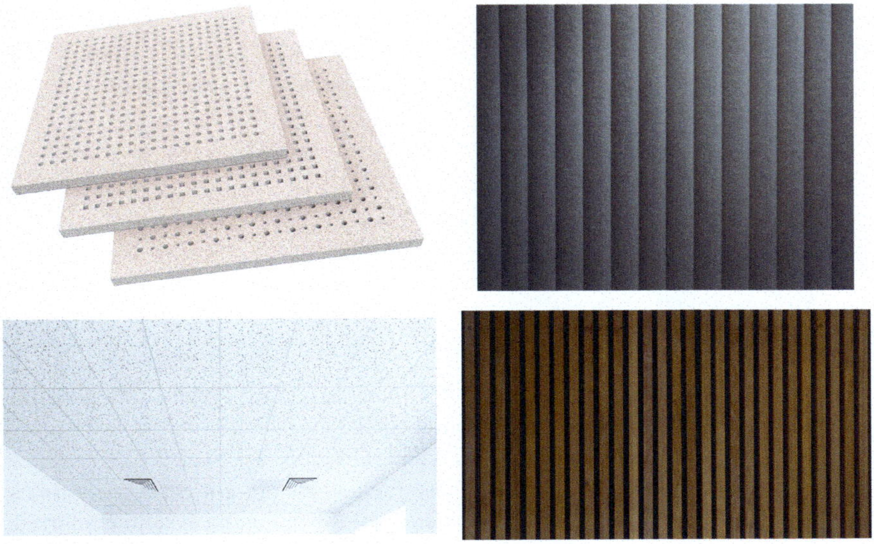

Fig. 8.7 Selection of porous absorbers

the absorption characteristics of the insulation is not noticeable. Covering a porous material with light fabric also provides a more attractive facing while having minimal effect on the absorption of the material, provided the flow resistance of the material is low.

Perforated Facing

A very common arrangement is to cover a porous absorber with a facing of solid perforated material, which may be timber, plaster, metal or any other suitable material. When the percentage open area of the perforations is less than 20%, the perforations may be designed to act as Helmholtz resonators (see below) and provide some additional low-frequency attenuation. However, this will be at the expense of high-frequency performance, which in most spaces is more important (for example, for control of sound from speech). Most such panels are designed with an open area of at least 20%, in which case the absorption of the panel is controlled by the properties of the underlying porous material.

Figure 8.7 shows a random selection from the many types of porous absorbing panels.

Diaphragmatic, or "Panel", Absorbers

Some of the sound energy striking a flexible panel mounted over an air space may be converted into vibrational energy, thereby reducing the amount of reflected sound.

Such "panel absorbers" are most effective at the frequency corresponding to the first resonance of the panel/air space system, where a high level of absorption can be achieved over a relatively narrow frequency range. An approximation to the resonance frequency is

$$f_r = \frac{60}{\sqrt{md}} \tag{8.8}$$

where

- m is the surface density of the panel in kg/m^2.
- d is the depth of the air cavity in metres.

Although some attenuation is achieved with an undamped system, this is significantly enhanced by placing sound absorption behind the panel, which prevents reverberant build-up of sound in the cavity. (Bies and Hansen 2009), Sect. 7.8 gives details of the expected attenuation. However, these predictions are not necessarily robust and the achieved attenuation depends on details of the mounting system employed.

Panels designed exclusively as diaphragmatic resonators are typically plywood or similar light material. For example, with an air gap of 200 mm, 12 mm plywood gives a resonant frequency of 46 Hz. These are typically used in high-end music spaces to provide absorption at low frequencies that is unachievable using porous absorbers. They are also used as "bass traps" in smaller music-oriented spaces such as recording studios (Fig. 8.8).

Some panels that primarily function as porous absorbers can be spaced away from a solid panel, giving additional low-frequency absorption. An example is ceiling tiles hung below the soffit, where the difference in low-frequency performance is demonstrated in Fig. 8.5.

Cavity, or Helmholtz, Absorbers

A cavity or Helmholtz absorber comprises an enclosed volume of air with a small neck opening. The energy of an incident sound wave is dissipated by the resonant vibrations of the air within the neck and enclosed volume, leading to high levels of sound absorption over a very narrow frequency range.

Dedicated Helmholtz absorbers are rarely used in building acoustics due to the limited frequency range over which they operate, although this can be extended somewhat by adding damping in the cavity. Their main use is to compensate for room modes in small music-oriented spaces.

Fig. 8.8 Diaphragmatic absorber

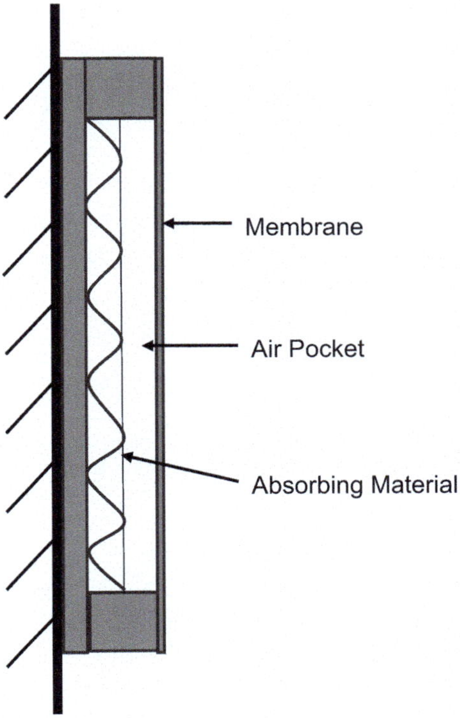

Microperforated Panel Absorbers

This form of absorber has been known and studied since the 1970's, but has recently become more popular as less expensive manufacturing methods have become available. It typically consists of a metal or plastic panel with submillimetre-sized holes and a perforation ratio of a few percent, which is spaced away from a solid surface. The void behind the plate may be empty or may be partitioned into cells or more complicated arrangements. The absorption coefficient for a basic plate, which depends on the plate thickness and hole diameter, is known, but the effect of various backing geometries is still the subject of research. (Herrin et al, n.d.) provides a good summary of their properties. Figure 8.9, adapted from (http://www.sandv.com/downloads/1712herr.pdf), shows an example of achievable absorption coefficients.

Comparison with Fig. 8.6 indicates that these absorbers are less effective at high frequencies than traditional porous absorbers but potentially more effective at low frequencies. Their main advantages for building acoustics are aesthetic—the perforations are almost invisible—and the absence of fibres. However, they remain significantly more expensive than porous absorbers.

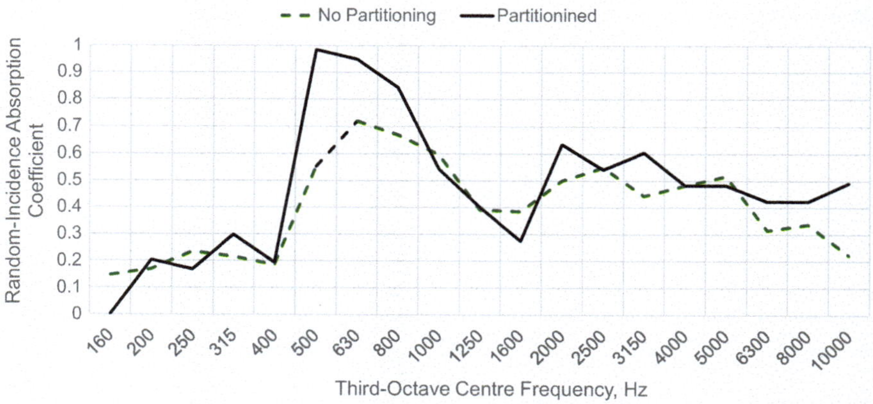

Fig. 8.9 Random-incidence absorption coefficients for a microperforated panel with and without partitioning of the void space. Adapted from (http://www.sandv.com/downloads/1712herr.pdf)

Reflectors and Diffusers

As discussed in Sects. 8.4 and 8.6 of the main text, when a reverberant environment is desired, it is often beneficial for sound to arrive at the listener from a range of directions. This can be assisted by adopting an irregular room geometry, but more importantly by ensuring that reflective surfaces scatter the arriving sound in multiple directions. Surfaces that achieve this are known a diffusers.

A diffuser can be simply a building element whose surface contains multiple irregular changes in slope. The length scale over which the changes occur is significant—diffusion will be maximised when the length scale is similar to the wavelength of the sound. Thus, irregularities on the scale of 30 cm will result in diffusion for sound at about 1 kHz, while irregularities on the scale of 1 m will result in diffusion at about 300 Hz.

In music performance spaces, diffusing surfaces are often placed at locations where echoes might otherwise occur (Sect. 8.4 of the main text).

Proprietary diffusing panels are available, although there is usually nothing special about these panels other than that they provide reflections in various directions—see Fig. 8.10.

For some specialist spaces, diffusers can be designed which give maximal diffusion over a wide range of frequencies. These are known as "quadratic residue diffusers" (Cox and D'Antonio 2006). Their use is generally restricted to recording studios and other very high-end audio applications.

Fig. 8.10 Diffusing panels

References

Standards Australia (2016) Standards New Zealand, AS/NZS 2107:2016 acoustics—recommended design sound levels and reverberation times for building interiors

Arau-Puchades H (1988) An improved reverberation time. Acoustica 65:163–180

Cremer L, Muller H (1978) Room acoustics (Translation by T. Schultz). Peninsular Publishing

Vercammen V (2010a) Sound reflections from concave spherical surfaces. Part I: wave field approximation. Acta Acoustica United Acoustica 96:82–91

Vercammen M (2010b) Sound reflections from concave spherical surfaces. Part II: geometrical acoustics and engineering approach. Acta Acoustica United Acoustica 96:92–101

Hopkins C (2012) Sound insulation. Elsevier

Sorbonne Universite. Audio/video anechoic database, [Online]. Available: http://www.lam.jussieu.fr/Projets/index.php?page=AVAD-VR

Ochmann M, Mechel F (F) Analytical and numerical methods in acoustics. In: Mechel F. Formulas of acoustics, Springer

Vorlander M (2008) Auralization. Springer

International Standards Organisation (2003) ISO 354:2003 acoustics-measurement of sound absorption in a reverberant room, International Standards Organisation
Sauro R, Mange G (2009) Absorption coefficients-part 2: is the edge effect more important than expected?, In inter-noise 2009, Ottowa
Bies D, Hansen C (2009) Engineering noise control, 4th edn. Spon Press
Herrin D, Liu W, Hua X, Liu J. A guide to the application of micro perforated panel absorbers, [Online]. Available: http://www.sandv.com/downloads/1712herr.pdf
Cox T, D'Antonio P (2006) A breief history of room acoustic diffusers. Acou Today 7:18–25

Chapter 9
Speech Intelligibility

Abstract This chapter describes the factors influencing speech intelligibility, which include human factors such as articulation and technical factors such as reverberation. Various measures of intelligibility are described, with a focus on the speech intelligibility index (STI), which is both predictable and measurable and has an established relationship with word recognition. Optimising acoustic performance for speech intelligibility is discussed, with a summary of good design practice. A spreadsheet is provided to estimate STI in simple cases.

9.1 Introduction

9.1.1 What Is Intelligibility?

Speech intelligibility represents how comprehensible speech is in an environment; in other words, how readily a listener can discern words and recognise them. Speech is rarely completely unintelligible or completely intelligible; there is a continuum between these two extremes.

- At the lower end of the intelligibility scale, the speech may be audible but none of it is understandable.
- At a slightly higher level, the intelligibility might be rated as "fair to poor". Here words are often unable to be recognised. This can occur in situations such as railway stations and transport terminals, retail plazas and airports, and may even occur in spaces such as courtrooms.
- At a higher level again, the intelligibility might be rated as "satisfactory". In this situation, the degree of word recognition by a listener depends on the context of the sentence, and it is likely that unfamiliar words may not be recognised.

Supplementary Information The online version contains supplementary material available at https://doi.org/10.1007/978-981-97-9380-8_9.

R. Bullen, *The Architecture of Sound*, Springer-AAS Acoustics Series,
https://doi.org/10.1007/978-981-97-9380-8_9

- At the upper end of the scale is the situation in which speech sounds completely natural and it is effortless for listeners to understand words.

When word recognition becomes easy, listeners are more able to concentrate on the subject matter. For example, recent research by Bild et al. (2021) with juries has shown that the clearer and more comfortable the speech is, the more easily juries are able to form an opinion because they do not miss key items that have been spoken.

Speech is often conveyed with the assistance of a sound reinforcement system, and in this chapter three types of communication are considered:

Type 1: The talker and listener are in the same acoustic space and the talker's voice is unamplified.

Type 2: The talker and listener are in the same acoustic space and the talker's voice is amplified by a sound reinforcement system. In this case the microphone, loudspeakers and room form the transmission channel.

Type 3: The talker and listener are in different acoustic spaces and the talker's voice is conveyed to the listener by a transmission channel comprising the talker's room and microphone, the electronic transmission or broadcast system, the loudspeaker and the receiver's room.

Speech intelligibility is fundamentally a qualitative topic as it involves human perceptions. It is true that each of the factors that affect intelligibility can be described quantitatively, as discussed below. However, it is important to also understand the human side of intelligibility, as this provides a context for the technical aspects. The factors that affect intelligibility are both human and technical, and loss of intelligibility is almost always due to a combination of both.

9.1.2 Speech Is a Set of Modulations

The speech signal is rich in temporal and frequency patterns, which can be regarded as fluctuations in both time and frequency. These fluctuations are called modulations.

When we say "aaaahhhh", our sound is relatively steady-state, but when we speak, we modulate a steady sound to produce syllables and words. Modulations are applied by the mouth and tongue to sounds produced by the vocal folds. The modulations occur largely over the frequency range 0.1 Hz to 20 Hz. If the temporal pattern of the modulations is disturbed, the intelligibility of the speech becomes degraded.

Figure 9.1a represents sound pressure variations for the spoken word "quip", after passing through a 1000 Hz octave-band filter. Figure 9.1b shows the "envelope" of the signal in Fig. 9.1a (the squared signal after passing through a low-pass filter), showing fluctuations occurring at frequencies of a few Hz. These fluctuations can be thought of as modulations imposed on a carrier signal, such as occurs in an AM radio broadcast. In the case of speech, there are numerous carrier signals. The speech can be considered in octave bands ranging from 125 Hz to 8 kHz, and each of these frequency bands will have its own set of modulations.

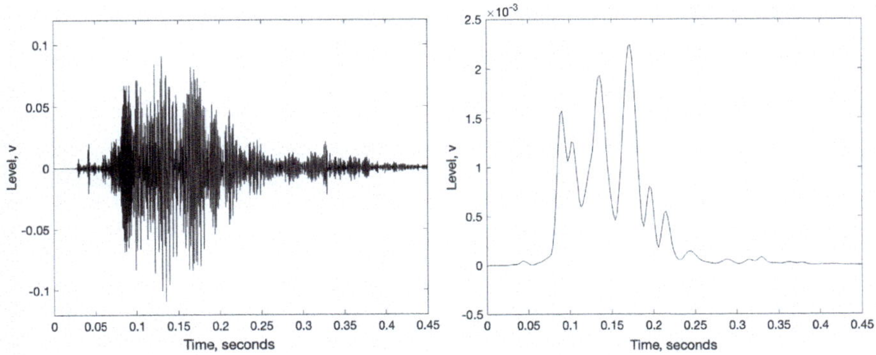

Fig. 9.1 **a** Temporal pattern of sound pressure in the 1000 Hz band occurring in the word "Quip" **b** "Envelope" of modulations in the 1000 Hz band

Modulations such as those in Fig. 9.1b will have a spectrum, with a maximum in the region 1–10 Hz. The "modulation index" is a measure of modulation strength. As defined for speech intelligibility assessment (Houtgast and Steeneken 1985), it is derived by:

- filtering the "envelope" function into low-frequency third-octave bands, typically 0.4 Hz to 20 Hz,
- normalising the RMS pressure in each band by the mean value of the envelope function.

Figure 9.2 shows an example of the spectrum of speech modulations derived from 30s of normal speech in the 1 kHz octave band.

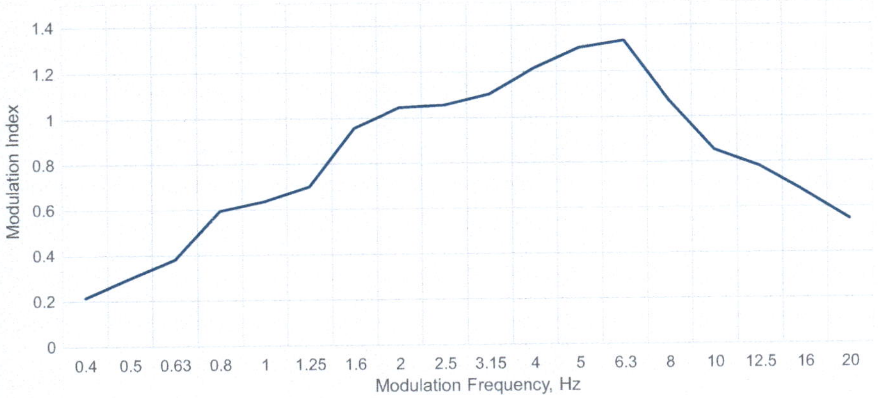

Fig. 9.2 Example of the spectrum of speech modulations in the 1 kHz octave band

Despite their acoustic complexity, spoken words can remain intelligible after drastic degradations in either the amplitude or frequency of modulations. However when reductions in the modulation index reach a certain level, words begin to be lost by the listener and sentences become unintelligible.

9.1.3 Masking and Self-masking

Sound from speech in certain frequency bands can be masked by ambient sound. This is discussed in Chap. 6, and particularly in Appendix to that chapter, in the context of masking of an intruding noise by background sound. In that case masking will be at a maximum for frequencies where the level of ambient sound approaches or exceeds that of the speech sound.

Speech can also be masked by other components of the speech itself, occurring at different frequencies or times. This is known as self-masking.

There are three types of self-masking mechanisms that occur in the ear and brain:

- upward masking—lower frequency sounds can mask quieter higher frequency sounds. This is the dominant self-masking mechanism, and a model of this mechanism is used in the MP3 compression algorithm to remove frequency content that is masked, to reduce the file size.
- downward masking—higher frequency sounds can also mask quieter lower frequency sounds, although this is not as significant in speech.
- temporal masking—this occurs when a signal and a masker are not presented simultaneously but are close in time. The masking sound may occur either before or after the sound that is masked.

Self-masking applies to individual words, and frequency-related masking depends on the spectrum of the word. Figure 9.3 shows a third-octave spectrum for the word "quip", assuming its overall level is 65 dBA L_{Aeq}. Also shown is the masking threshold due to upward masking from sound in the 500 Hz third-octave band. (See Chap. 6, Appendix for calculation of these thresholds.) It is clear that sound in the adjacent band is very nearly masked by sound in the 500 Hz band.

This situation is common and demonstrates that any process that applies a low-pass filter to speech sound risks making some sound unintelligible simply through self-masking.

9.1.4 Speech Levels and Octave-Band Importance

The importance of a particular speech frequency band for speech intelligibility is complex and depends on the extent to which the band provides information that is redundant or complementary to that of other speech bands.

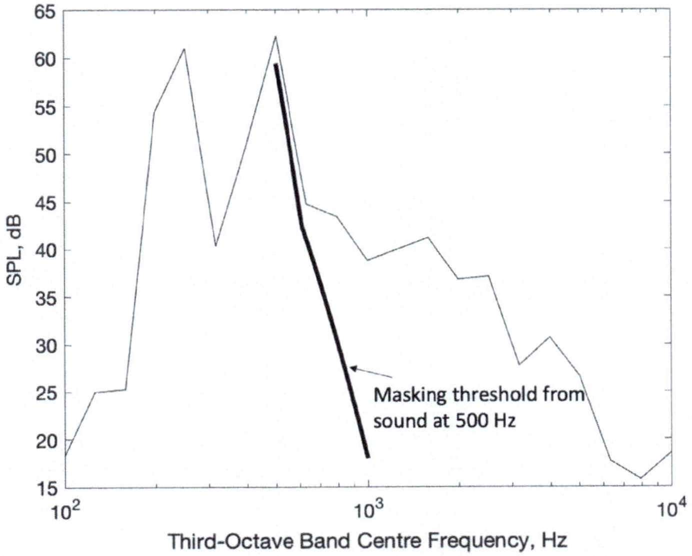

Fig. 9.3 Third-octave spectrum of the word "quip", showing the masking threshold due to sound in the 500 Hz band

L_{eq} values of typical speech at 1 m, in third-octave bands, are given in the Supplementary file1 spreadsheet in Chap. 2. These are taken from ASA/ANSI Standard S3.5-1997 (Acoustical Society of America 2020) and represent an average for male and female speech. (Other sources report slightly different values, notably (International Electrotechnical Commission 2020) which shows a spectrum for male speech).

Table 9.1 shows octave-band levels derived from the values in the Supplementary file1 spreadsheet for normal speech directly in front, together with estimates of the relative importance of each band for understanding speech. These "importance" values are based on "weighting factors" used in the calculation of the speech transmission index (STI), which is described in Sect. 9.5.1 (International Electrotechnical Commission 2020). (Although the quoted values apply specifically for male speech they give an indication of importance overall.) The most important bands are 500 Hz to 8 kHz, with maximum importance at 2 kHz.

Table 9.1 Typical octave-band sound pressure levels (in dB L_{eq} at 1 m) for normal vocal effort directly in front of the talker, and importance values assigned in calculation of STI

	Octave band frequency (Hz)						
	125	250	500	1000	2000	4000	8000
L_{eq} @ 1 m, normal voice	51	57	59	54	49	44	39
"Importance", based on weighting factors assigned in STI	0.085	0.127	0.230	0.233	0.309	0.224	0.173

Exercise 9.1

Based on Table 9.1, find the direct (that is, free-field) L_{eq} level of a normal voice at a distance of 2 m, in octave bands.

(a) Assume the background noise level follows the NR40 curve. (NR curves are in the Supplementary file1 spreadsheet.) In what frequency bands is the background noise level higher than the L_{eq} level of speech at 2 m?
(b) What are the importance levels of these bands, from Table 9.1?
(c) Repeat (a) if the background level follows the NR45 curve.
(d) Now what are the importance levels of bands where the background level is higher than the speech L_{eq}?

Now assume that the speaker turns their head by 90 degrees. Use the directionality data in the Supplementary file1 spreadsheet to estimate the octave-band L_{eq} from their voice at 2 m. [You will need the difference between 0 and 90 degrees levels, and you will need to estimate at 125 Hz].

(e) If the background sound follows the NR40 curve, what are the importance levels of frequency bands where the background level is higher than the speech L_{eq}?

Exercise 9.2

The diagrams below show third-octave spectra for three words. Which of these words would be most likely to suffer from self-masking? (Fig. 9.4).

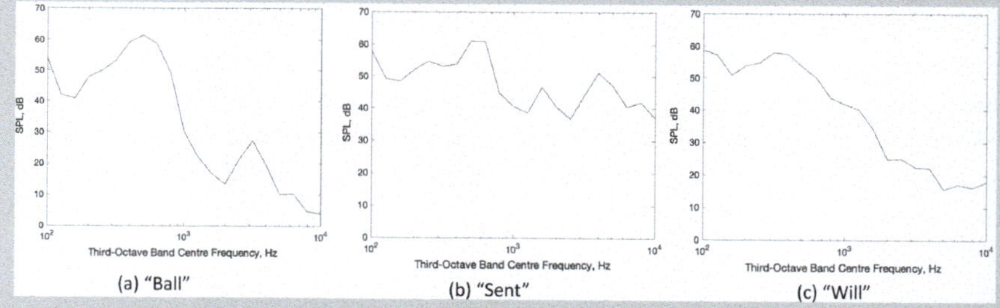

Fig. 9.4 Spectra for three words

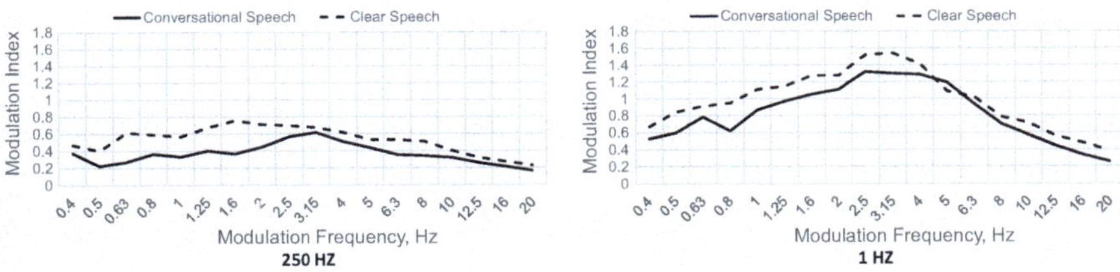

Fig. 9.5 Examples of modulation spectra in two octave bands for clear speech and conversational speech—adapted from (Krause 2001)

9.2 Human Factors Affecting Intelligibility

9.2.1 *Talker Articulation and Word Delivery Rate*

We all know that if someone mumbles it can be difficult to understand what they are saying. Sometimes, no matter how the speech is articulated, if the word-delivery rate is high, we struggle to understand. Both the intensity level and spectra of the modulations depend on the degree of articulation and the word-delivery rate. Figure 9.5, based on data from (Krause 2001), shows examples of modulation spectra for conversational speech and for speech delivered clearly, but at the same rate. Clear speech involves greater modulation, particularly at lower modulation frequencies.

9.2.2 *Accent Differences*

Talkers and/or listeners who are non-native English speakers have an inherent disadvantage due to accent differences, which can produce a significant loss of intelligibility. Given the huge range of accents that exist in English, some accent differences can cause difficulties equivalent to those of a non-native speaker.

9.2.3 *Hearing Loss*

Age-related hearing loss affects an increasing proportion of the community. Currently, 15% of the adult population suffers from some form of hearing loss, and this is expected to increase to 22% in 2060. Hearing loss not only places some components of the speech signal below the threshold of hearing, it is also generally higher at higher frequencies, exacerbating self-masking.

9.3 Technical Factors Affecting Intelligibility

There are four primary technical factors that affect intelligibility—ambient noise, reverberation, echoes, and spectral fidelity. Each factor can be considered as a "noise" that is imposed onto the pure original speech. As such, the degradations can be considered on a signal-to-noise basis.

9.3.1 Ambient Noise

Noise may cause masking of speech at all frequencies, depending on the spectrum of the noise.

In Type 1 situations (unamplified listening in the same space), the talker and the listener are likely to be exposed to a similar level of noise and the speech-to-noise ratio is set by the ambient noise in the space.

However, In Type 2 and 3 situations, the listener could be in a quiet situation, while the talker could be in a noisy environment. The ambient noise at the talker's microphone will then be transmitted to the listener, along with their voice.

Ambient noise is often transient—for example, trains arriving at and departing from a railway station platform. Hence in some cases the signal-to-noise ratio should be considered both on a long-term basis where ambient noise is described by the L_{eq} metric, and on a worst-case basis using the maximum level.

Figure 9.6 illustrates the mechanism by which ambient noise reduces the modulation in speech in any frequency band. Assuming that ambient noise will be approximately constant on time scales of a few seconds, this noise affects the modulations in speech equally at all modulation frequencies (at the same carrier frequency). Essentially, the noise fills in the sections of the modulation that have low levels.

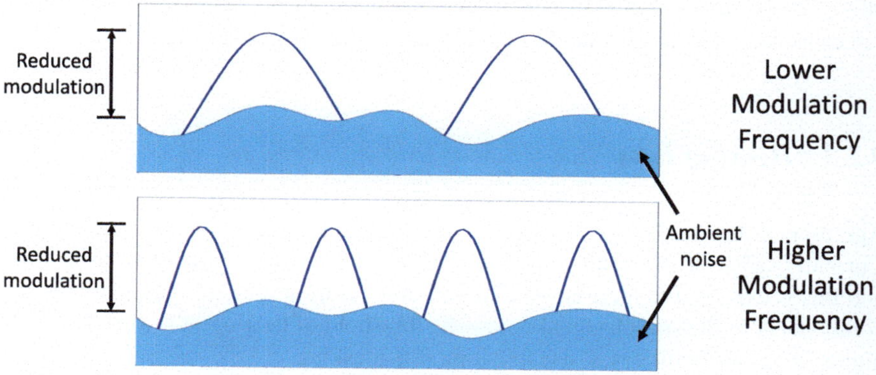

Fig. 9.6 Conceptual illustration of the mechanism by which ambient noise reduces the modulation equally at all modulation frequencies

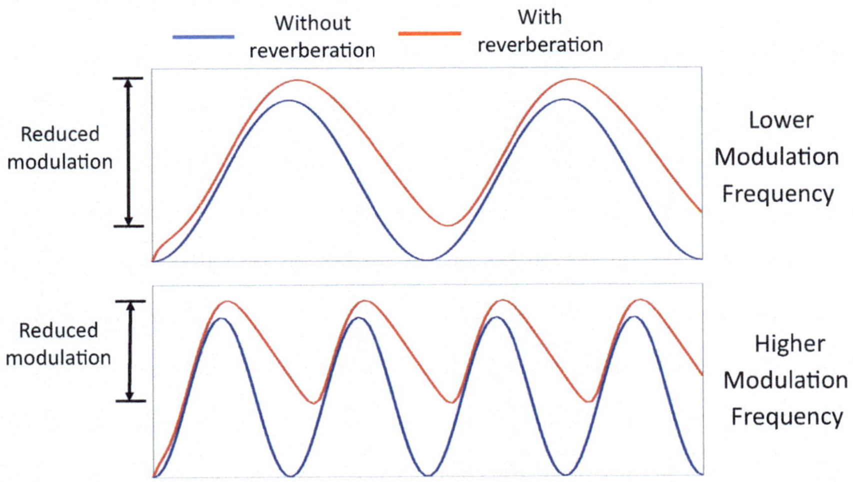

Fig. 9.7 Conceptual illustration of the way that reverberation reduces the level of the modulation at different modulation frequencies

9.3.2 Reverberation

Reverberation creates a time-based smearing of the speech sounds and reduces the level of the modulations in a frequency-dependent manner. When a particular speech sound such as a vowel is reverberating and dying away, a subsequent softer sound such as a consonant can be masked by the reverberation of the earlier sound. The longer the reverberation time in a space, the more masking occurs.

Figure 9.7 illustrates the way that with a given reverberation time, low-frequency modulations are "filled in" less than higher frequency modulations. When the reverberation decay over the period of a modulation waveform is small, the modulation depth will be significantly reduced.

The loss of modulation caused by reverberation can be considered as an unwanted signal, i.e. self-induced "noise". The ratio of the level of un-reverberated speech at each frequency to the level of reverberated speech determines the operative signal-to-noise ratio.

9.3.3 Echoes

Echoes are strong reflections that arrive after the direct sound arrives, and they add replicas of the original speech which can mask speech sounds (see Sect. 8.4). Accordingly, echoes can also be considered as self-induced noise.

If the arrival time of the echo relative to the direct field, ΔT, is sufficiently small, the modulations fuse together, and the modulations in the echo add to the direct-field

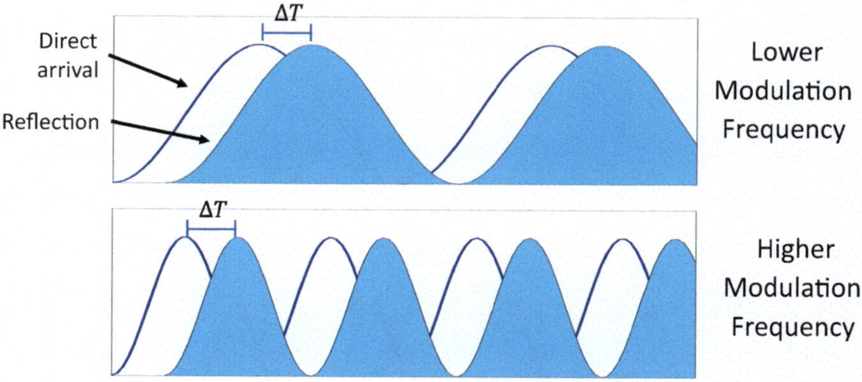

Fig. 9.8 Conceptual illustration of the way that echoes reduce the level of the modulation depending on the difference between the arrival times of the original signal and the echo

modulations. However, if ΔT is large, then the echo will fill in the trough of the direct-field modulation. Figure 9.8 illustrates this effect.

Sound systems using multiple loudspeakers to cover a wide listening area can also result in echoes due to the late arrival of direct sound from distant loudspeakers.

9.3.4 Spectral Fidelity of the Speech

The fourth major technical factor that affects speech intelligibility is the spectral balance of the speech received by the listener.

In Type 1 situations, degradation of the frequency response arises from characteristics of the space and is described as "colouration" in Sect. 8.5. In Type 2 and 3 situations it is also due to the frequency response characteristics of the electroacoustic system used. This degradation can increase masking in two ways.

i. Additional Masking by Ambient Noise

Most obviously, the transmission channel may reduce the level of the speech signal at some frequencies to the extent that it is masked by ambient noise or, more rarely, is below the threshold of hearing. This occurs primarily with Type 2 and 3 transmission channels, which often contain a strongly acting low-pass filter. (An example is the sound produced by car radios tuned to the AM band. These radios typically have a low-pass filter with a turnover point around 3 kHz.)

ii. Additional Self-masking

When the transmission channel has a reduced high-frequency response, the potential for self-masking is also increased. This can occur in all types of transmission.

Currently, in the design of rooms and systems for which speech is important, spectral fidelity is often not given due consideration. This situation is unfortunate as spectral fidelity can be as important for intelligibility as the control of reverberation.

9.4 Optimising Intelligibility

Poor articulation, rapid word-delivery rate and accent differences for the speaker, as well as hearing loss for the listener, can substantially degrade speech intelligibility even without considering properties of the speech transmission channel. A design that is robust against these factors needs to include minimal additional degradation from the transmission channel.

To optimise intelligibility, the following factors must be addressed:

- ambient noise,
- reverberation characteristics, in particular the ratio of early-arriving sound to late-arriving sound,
- frequency response of the transmission channel.

9.4.1 Ambient Noise

In meeting rooms and similar spaces where speech intelligibility is critical, the background noise must be suitably low and ideally not more than 35 dBA.

In these rooms, an ambient noise level of 40 dBA or more can result in a significant reduction in intelligibility with a softly spoken talker facing away from a listener. This is exacerbated in remote conferencing situations, due to other factors in the transmission channel and the listening room, and the reproduced speech at remote listeners may well be poor.

AS/NZS 2107:2016 (Standards Australia; Standards New Zealand 2016) recommends a range of background levels from 30 to 40 dBA L_{Aeq} for "board and conference rooms", 30–35 dBA for "court rooms" and ranges up to 40 dBA in other similar spaces (see Chap. 5). Levels at the upper end of these ranges will not provide good acoustic comfort and robust intelligibility for speech under non-ideal conditions.

9.4.2 Reverberation

For optimum transmission of speech from a talker to a listener, both the listening and talking environments should have reverberation times that are suitably low and constant with frequency. As noted in Chap. 8, reverberation times suggested for various spaces in AS/NZS 2107 should be considered in the light of the specific requirements for each space. In particular, for good speech intelligibility, lower values than those suggested in the standard are often required.

Further, reverberation should be considered in three dimensions. As noted in Chap. 8, the simple Sabine or Norris-Eyring equations used for reverberation predictions assume that sound absorption is evenly distributed over all surfaces in the space. When absorption is not evenly distributed, late reverberation decays more slowly, causing a decrease in modulation depth.

For optimal intelligibility, rooms should have sound absorption fitted to all surfaces in a distributed manner. If this cannot be achieved, then sound absorption should be on two orthogonal surfaces, for example a side wall and an adjoining end wall. The most problematic arrangement is one with parallel non-absorptive surfaces, allowing reverberation to occur between the surfaces which dominates the slope of late reverberation in the room.

This potential problem can be identified by including reverberation time predictions using a 3D equation (such as the Fitzroy or Arau-Puchades equations described in Chap. 8) to highlight potential excessive late reverberation associated with unequal distribution of absorption.

In an attempt to create clean lines and hard surfaces, many rooms where speech intelligibility is critical create excessive reverberation. This trend is also promoted by factors discussed in Chap. 10. Even when intelligibility within these rooms is adequate, if sound is transmitted to any remote participants it is likely to be poor.

9.4.3 Early-to-Late Ratio (Clarity)

The above discussion implies that reverberation will always degrade intelligibility, which is not quite accurate. Reflections that arrive at the listener within a specific time window of the arrival of the direct field, called the early-time window, are usually beneficial for intelligibility, as human hearing fuses those reflections with the direct field. This produces an effective increase in the direct-field level, thereby increasing the ratio of the signal to ambient noise and later arriving reverberation. The early-time window is generally taken to be in the range of 35 ms to 50 ms.

Sound arriving after the early-time window is called late-arriving sound, and consists of reverberation and echoes which are detrimental to intelligibility. The early-to-late ratio for speech is called C_{50} or C_{35} depending on the length of the early-time window. Here "*C*" refers to "clarity". The C_{50} index is also described in Chap. 8 in the context of sound quality. It is defined as

$$C_{50} = 10\lg\left(\frac{E_{\mathrm{direct}}+E_{\mathrm{early}}}{E_{\mathrm{late}}}\right), \tag{9.1}$$

where

- E_{direct} is the direct sound energy,
- E_{early} is the cumulative reflected energy up to 50 ms after the direct sound arrival,

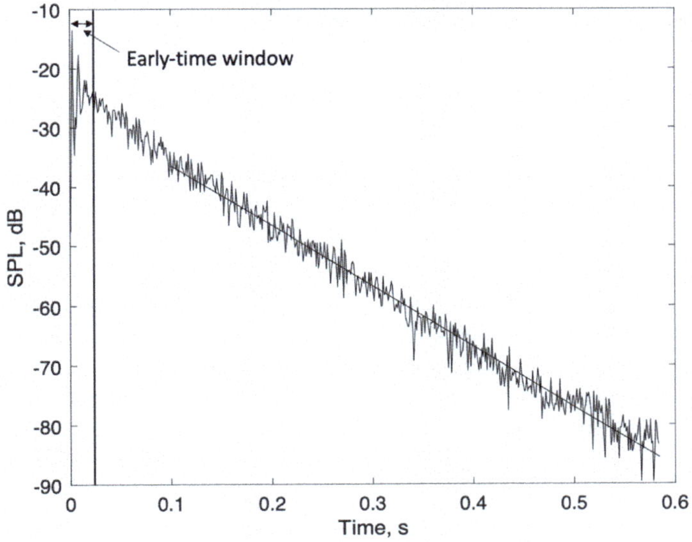

Fig. 9.9 Reverberation trace in a room, showing the early-time window

- E_{late} is the cumulative reflected energy after 50 ms.

  This is illustrated in Fig. 9.9.
  Clarity can be maximised by

- minimising reverberation (specifically late reverberation) as described in the previous section, as well as late echoes, and/or by
- maximising early reflections.

Achievable values will depend on the situation, but a value of 6 dB for C_{50} can be considered good, with 9 dB being seen as a goal.

Sound reinforcement systems can increase clarity by using loudspeaker directivity to concentrate early sound at the listener position. These designs are generally used with a computer model to optimise the position and characteristics of the loudspeakers.

9.4.4 Frequency Response of the Transmission Channel

Control of the spectral response of a room is described in Sect. 8.5 and involves consideration of room modes and the absorption properties of materials in the room. In Type 2 and Type 3 situations, the frequency response of all electrical and electroacoustic elements of the transmission chain also needs to be specified.

The aim is to achieve an overall frequency response that is as flat as possible in the important speech frequency range of 500 Hz–8 kHz.

9.5 Measurement of Speech Intelligibility

9.5.1 Metrics

The only accurate way to measure speech intelligibility is to undertake a word-recognition test, commonly called a word-score test. This involves a talker reading a series of nonsense words, with the listener writing down the words that they hear. As this type of test takes a long while to execute, it is impractical in most situations. However, word-score results can take account of all factors affecting intelligibility, including poor articulation, accent differences between talker and listener's native languages, hearing loss, ambient noise and reverberation/echo.

It is much quicker to measure the degradation of the transmission channel only, assuming that the speaking and hearing sections of the channel are perfect. A transmission channel can be as simple as a room, or as complex as a voice in one room being transmitted via the internet to a listener in another room.

Metrics that have been proposed to assess the performance of a speech transmission channel include the following.

i. Speech Interference Level

The speech interference level (SIL) is not an intelligibility metric per se, but allows estimation of the vocal effort that would be required when speaking in a room with ambient noise. SIL is computed as the arithmetic average of the background sound pressure levels in the four octave bands, 500 Hz, 1000 Hz, 2000 Hz and 4000 Hz.

Figure 9.10, based on (Bies and Hansen 2009), shows the vocal level that would be required for just-adequate speech communication at various distances from the talker

ii. Articulation Index

The articulation index (AI) has been used for a long while and addresses degradation due to ambient noise only. The AI metric simply computes the difference in each 1/3 octave band between the long-term RMS levels of the speech and interfering noise. Naturally, AI assumes that the noise is steady state. In recent times, AI has been replaced by the speech intelligibility index metric which is more sophisticated.

iii. Articulation Loss Of Consonants

This is an old metric pertaining only to the 2 kHz band and should not be used.

iv. Speech Intelligibility Index

The speech intelligibility index (SII) is a measure of a transmission channel's ability to provide speech intelligibility. It ranges between 0.0 and 1.0 and is described in ANSI, S3.5, 1997 (R2017) (Acoustical Society of America 2020). In contrast to the speech transmission index (STI) described below, SII is calculated and cannot be measured directly. A number of methods are described in (Acoustical Society of

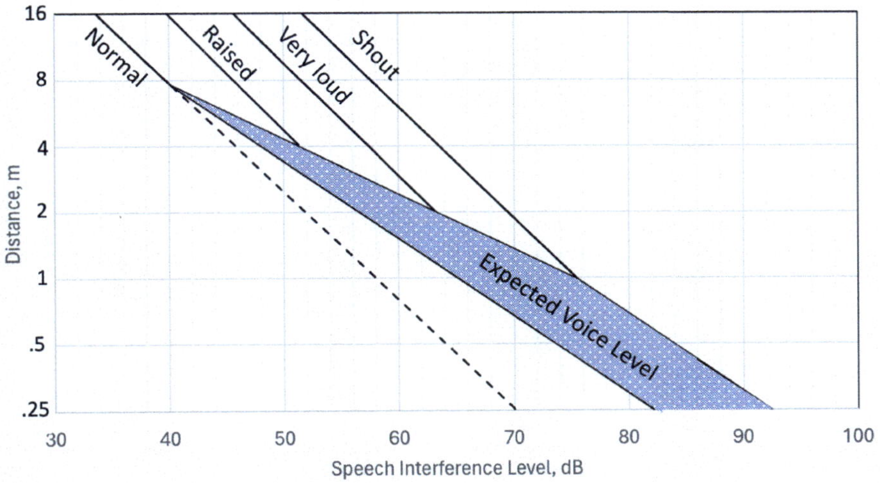

Fig. 9.10 Vocal level required for just-adequate speech communication at various distances from the speaker, adapted from (Bies and Hansen 2009)

America 2020) to derive the various components SII for various situations, using measurements and/or standard data.

SII is commonly used by experimental audiologists, because of its higher frequency resolution and its sensitivity to the intelligibility decrease at high vocal efforts. SII does not appear to be significantly more sensitive than STI to the effects of frequency response.

v. Speech Transmission Index

The most common metric for the loss of intelligibility in a transmission channel is the speech transmission index (STI). This test method is widely used in Australia and Europe and measures the loss of modulation occurring in the transmission channel. The STI method is described in the standard IEC 60268-16-2020 (International Electrotechnical Commission 2020). That standard provides a substantial description of the various processes and equations contained within the process.

STI measures the loss of modulation for fourteen modulation frequencies contained in each of seven frequency bands from 125 Hz up to 8 kHz. Modulation frequencies cover the range 0.63 Hz to 12.5 Hz, creating a matrix of 14×7 different modulations, each of which is measured separately. The loss of modulation for each cell of the matrix is described by the modulation transfer function (MTF).

The MTFs can then be adjusted for auditory masking and threshold effects, and combined to produce a modulation transfer index for each octave band. These are combined in turn, to form a final STI which is reported on a scale from 0.0 to 1.0, where 0.0 is completely unintelligible. Details of this procedure are given in IEC 60268-16-2020 (International Electrotechnical Commission 2020).

STI accounts for the following factors:

- ambient noise level,
- reverberation and echo,
- high- and low-pass filtering,
- duration,
- upward masking (in the human ear),
- threshold of hearing.

It is important to note that STI cannot properly consider the effects of poor frequency response of the transmission channel when the ambient noise is low. In addition, STI is unable to properly account for the effect of long echo delay times on intelligibility.

9.5.2 Measuring STI

There are two methods by which STI can be measured—direct and indirect.

i. Direct Method

The direct method employs a special test signal that contains the 98 modulations over the seven octave bands, which is broadcast into the transmission channel. A sound level meter fitted with special software receives the signal and measures the loss of modulation in each of the 98 cells of the MTF matrix, and STI is computed from these.

The L_{eq} of the ambient noise that will be present during operation of the transmission channel must also be measured or computed in each octave band, and introduced into the STI algorithm.

Until 2019, the time to measure the loss of modulation for all 98 modulations was fifteen minutes, which is too long for practical applications. To shorten the measurement time, a derivative of the STI method called STIPA was developed which employs a sparse set of fourteen modulations, two in each of the seven bands spanning 125 Hz to 8 kHz. Several versions of STIPA were formulated in different editions of (International Electrotechnical Commission 2020).

For measurements, ambient noise in the measurement environment must be steady without transient noise such as a door closing, dog bark, etc. Measurements in conditions of non-steady background noise result in erroneously high values of STI. If STI is to be assessed for a situation with noise that varies considerably, the STI should be measured in a low-noise situation and the equivalent operational noise mathematically added into the STI.

ii. Indirect Method

In the absence of direct measurements, the MTF matrix, and hence STI, can be computed from the impulse response of the transmission channel and the L_{eq} of the total "noise" heard by the listener. The total "noise" is a combination of ambient noise and the effective noise due to masking and the hearing threshold.

Appendix 1 gives a detailed explanation of the formulae involved.

A simplified calculation procedure exists (see Appendix 1) which is valid only in spaces where a statistical reverberant field is present for almost all the decay—referred to here as "Sabine spaces". This method is unable to consider the effects of non-linear sound decay and of echoes. This simplified procedure is implemented in the Supplementary file1 spreadsheet.

Both the direct and indirect methods are able to measure the degradation in intelligibility due to ambient noise (steady state), reverberation and echo in the measurement space. However, the indirect method cannot accommodate the effects of non-linear distortion by electronic systems, such as signal clipping.

9.5.3 Relationship Between Word Scores and STI

The relationship between STI and speech intelligibility is illustrated in Fig. 9.11. The relationship shown was recently developed by Morales (Morales and Li 2018) and uses the so-called "Harvard list" of phonetically balanced (PB) words and binaural listening. The percentage of words heard correctly is related to STI, with degradation of the speech signal caused by reverberation, noise, bandpass filters and echoes.

Table 9.2, which is adapted from (International Electrotechnical Commission 2020), associates STI values with the type of messages that can be satisfactorily used and examples of use in the field. However, note that these recommendations are subject to the qualifications listed in Sect. 9.5.4.

The example environments described in Chap. 2, Appendix 2 provide a general indication of the sound of speech at various STI values. For each of the eight impulse

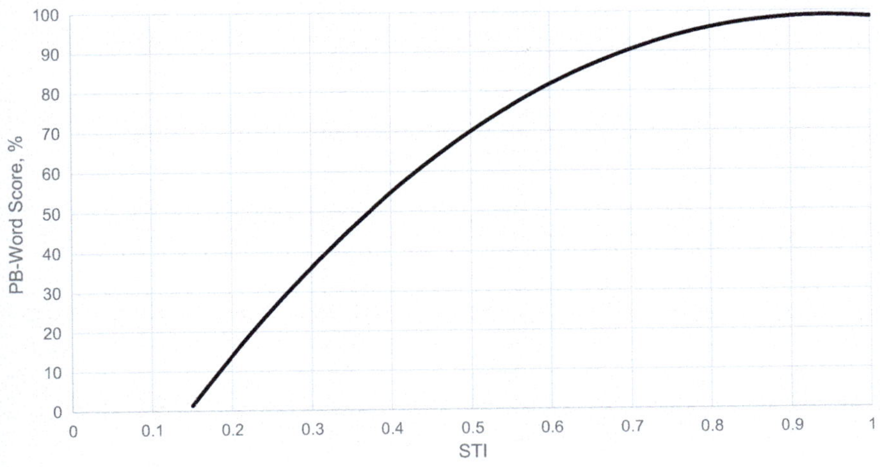

Fig. 9.11 Relationship between STI and the percentage of phonetically balanced (P-B) words heard correctly, from (Morales and Li 2018)

Table 9.2 Relationships between STI values and application (from International Electrotechnical Commission (2020))

Nominal STI value	Type of message information	Examples of typical uses (for natural or reproduced voice)	Comment
> 0.76		Recording studios	Excellent intelligibility but rarely achievable in most environments
0.74	Complex messages, unfamiliar words	Theatres, speech auditoria, parliaments, courts, assistive hearing systems	High speech intelligibility
0.7	Complex messages, unfamiliar words		
0.66	Complex messages, unfamiliar words	Theatres, speech auditoria, teleconferencing, parliaments, courts	High speech intelligibility
0.62	Complex messages, familiar words	Lecture theatres, classrooms, concert halls	Good speech intelligibility
0.58	Complex messages, familiar context	Concert halls, modern churches	High-quality PA systems
0.54	Complex messages, familiar context	PA systems in shopping malls, public buildings offices, voice-alarm systems, cathedrals	Good quality PA systems
0.5	Complex messages, familiar context	Shopping malls, public buildings offices, voice-alarm systems	Target value for voice-alarm systems
0.46	Simple messages, familiar words	Voice-alarm and PA systems in difficult acoustic environments	Normal lower limit for voice-alarm systems
0.42	Simple messages, familiar context	Voice-alarm and PA systems in very difficult spaces	Limited intelligibility
0.38		Not suitable for PA systems	–
< 0.36		Not suitable for PA systems	–

responses considered, STI was calculated for reverberation only, and without corrections for auditory masking and threshold. (These corrections are necessary in actual measurements but are not valid when there is no control over the listening level.) Table 9.3 shows these reverberation-only STI values, and simulated speech in these environments can be heard by clicking the name of each room.

Table 9.3 STI values (reverberation only) for example environments. (Open the associated Supplementary file to hear speech as heard in that space)

Space	STI
Conference Room [Link to Supplementary file3}	0.66
Meeting Room [Link to Supplementary file4}	0.77
Hotel Room [Link to Supplementary file5}	0.64
Lecture Room [Link to Supplementary file6}	0.66
Shared Office [Link to Supplementary file7}	0.70
Stairwell [Link to Supplementary file8}	0.36
Large Office [Link to Supplementary file9}	0.70
Concert Hall [Link to Supplementary file10}	0.49

9.5.4 *Shortcomings of the STI Metric*

As for any single-number metric, STI cannot fully describe the potential loss of speech intelligibility in all situations. This may be due to multiple factors, both human and technical. In particular:

(a) The recommendations in Table 9.2 and the word score values in Fig. 9.11 are based on the assumptions that talker articulation is excellent, that talker and listener have the same native language and accent, and that the listener has good hearing.

(b) Although word intelligibility scores give an overall measure of performance, certain words will be more likely to be misunderstood than others. In particular, self-masking becomes much more likely for words with soft consonants at the end such as "salve", "sheik", "quip" and "dung" due to the predominance of low frequencies in the spectrum up to 6 kHz.

c) No STI verification has been performed for speech in which the tonal structure between 250 Hz to 4 kHz has been significantly distorted (that is, the transmission channel does not have a "flat" frequency response). As such, the relationships in Fig. 9.11 may not properly represent the increased loss of intelligibly that occurs with self-masking when poor frequency responses causes higher-energy vowels to mask lower-energy consonants.

In summary, the speech transmission index is an extremely valuable tool to understand the ability of a speech transmission channel to provide intelligibility for listeners. However, it should not be relied upon as the only specification required to provide adequate speech intelligibility in a building design.

9.6 Summary of Good Design Practice

Where speech intelligibility is important for the functioning of a space, good design practice begins with consideration of the factors listed in Sect. 9.4. Recommended design procedures for speech intelligibility can be summarised as follows.

1. Consider background sound (as described in Chap. 5). Maximum values in the ranges specified in AS/NZS 2017:2016 may not be appropriate where speech intelligibility is important, and values toward the lower end of those ranges are recommended.
2. Consider the reverberation characteristics of both the listening space and (in the case of Type 3 situations) the speaker's space. This includes both the reverberation time (measured for example as RT_{30}) and the early/late ratio (measured as C_{50}). The latter is maximised when the decay is linear, which as described in Chap. 8 means maximising diffusion and distributing absorption roughly equally over all surfaces. With regard to RT_{30}, note once again that the values suggested in AS/NZS 2017:2016 may be too high where speech intelligibility is important.
3. Consider the frequency response of the transmission channel. In Type 1 situations this is the "colouration" of the reverberated sound, as described in Chap. 8. In Type 2 and Type 3 situations it also includes the frequency responses of microphones, electronic processors and speakers. The frequency response should be as flat as possible in the speech frequency range of 250 Hz–8 kHz.
4. After the above considerations, calculate STI for the space. Note that the simplified calculation procedure in Appendix 1 (and implemented in the Supplementary file1 spreadsheet) is appropriate only where the reverberation decay can be assumed to be linear. Otherwise, a full impulse response should be calculated using modelling software, for each relevant source and receiver location.
5. The word score associated with the calculated STI and the recommendations of Table 9.2 be considered, remembering the possibility of degradation due to human factors and technical factors not considered in STI. If the score is not considered appropriate, re-design is required.
6. If an appropriate STI has been achieved, this should be specified along with the design details required to achieve it. This allows for post-construction verification. However, achievement of the STI value in the absence of the other specified design details should not be considered acceptable.

Exercise 9.3

You are about to make a speech in a room where listeners will be up to 5 m away, and you wish to predict the STI at your listeners' position when you are speaking.

You visit the room when it is empty and measure its internal ambient noise level. The spectrum of the noise is essentially equal to the NR40 contour. The volume of the room is about 350 m^3.

(a) Based on the background noise described by NR40, use SIL (Fig. 9.10) to determine your likely vocal effort.
(b) Using the speech data in the Supplementary file1 spreadsheet in Chap. 2, find the expected sound power level of your voice in *octave bands* from 125 Hz to 8 kHz. (You will need to estimate at 125 Hz and 8 kHz.)
(c) You have estimated that the reverberation times will be 1s at all frequencies due to the hard wall, floor and ceiling surfaces. Use the sound power levels to compute the reverberant levels of speech in each octave band, assuming the room is a Sabine space.
(d) As the room is quite wide, listeners on one side of the room may be substantially off axis from your mouth. Assume that a listener will be 90 degrees off axis and use the data in the Supplementary file1 spreadsheet to estimate the directional loss at each octave-band frequency. Calculate the direct-field level at 5 m.
(e) Using the Supplementary file1 spreadsheet, which implements the procedure in Appendix 1, calculate the STI based on the following factors:

- Ambient noise of NR40
- Speech levels as determined in Steps (c) and (d)
- Assume that the cumulative energy of the early reflections at the furthest listener will equal the energy in the direct-field sound (that is, the "total direct field" has twice the energy of the "direct field" calculated above).

(f) Estimate the word loss and the type of message that could be reliably transmitted to all listeners.
(g) Determine the STI if the restaurant were to install:

- sound-absorption panels to reduce the build-up of patron-noise, yielding an RT of 0.5 secs at all frequencies and also
- treatment to reduce background noise levels by 3 dB at all frequencies.

(Remember that altering the reverberation time will also alter the level of the reverberant field.)

Appendix 1: Indirect Calculation of STI

Calculation of the Modulation Transfer Function

This procedure is set out in IEC Standard 60268-16 2020 International Electrotechnical Commission (2020).

Equation (9.2) below allows the modulation transfer function to be computed from:

- the impulse response of a transmission channel and
- the L_{eq} of the total "noise" heard by listener.

Impulse response is described in Appendix 2 to Chap. 2. The total "noise" is a combination of ambient noise and effects due to masking and the hearing threshold.

$$m_k(f_m) = \overbrace{\frac{\left|\int_0^\infty h_k(t)^2 e^{-j2\pi f_m t} dt\right|^2}{\int_0^\infty h_k(t)^2 dt}}^{\text{A}} \cdot \overbrace{\left[1 + 10^{-\text{SNR}_k/10}\right]^{-1}}^{\text{B}} \tag{9.2}$$

where

- modulation transfer function $m_k(f_m)$ is the transfer function for modulation frequency f_m in octave band k
- $h_k(t)$ is impulse response of octave band k
- SNR_k is the signal-to-noise ratio in band k in dB.

Part A of the equation computes the loss of modulation due to reverberation and echo.

Part B refers to the combined "noise" level consisting of ambient noise, the effective noise introduced by masking and the threshold of hearing. The combination of both parts gives the modulation transfer ratio at each modulation frequency in each octave band.

The complete modulation transfer function is then processed for redundancy and band-importance as described International Electrotechnical Commission (2020) to produce the STI value.

3D acoustic modelling software, using ray-tracing to generate impulse responses, computes the noise-free MTF using part A of Equation (9.2) only.

As the impulse response of a system is only defined for linear, time-invariant systems, the indirect measurement is unable to properly account the effects of non-linear distortion that may occur in electronic transmission, or of changes in the transmission path.

Examples of MTF Matrices with Reverberation and Echo

i. **Reverberation**

Figure 9.12 shows an impulse that has been reverberated with a reverberation time (RT_{30}) of approximately 2.1s in each octave band. The original impulse was not added in and therefore the impulse response does not have a "direct-field" component before the commencement of the reverberation.

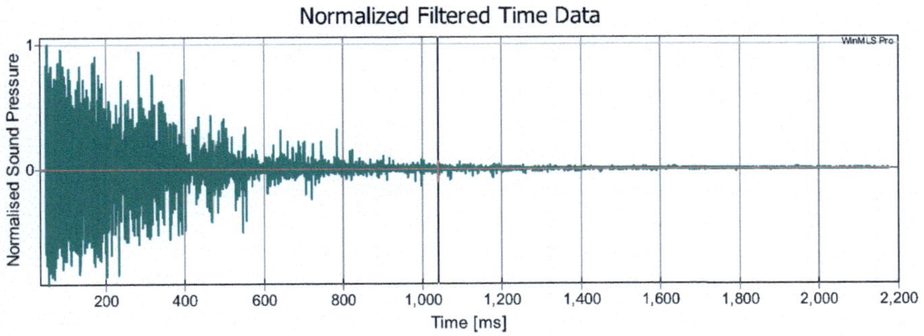

Fig. 9.12 Reverberated impulse with an RT30 of 2.1 s

Figure 9.13 presents a logarithmic version of the impulse response in which the y-axis is in decibels and normalised to 0 dB, (i.e. a maximum linear pressure of unity).

Table 9.4 shows the MTF matrix computed from this reverberated signal. The modulation ratios decay monotonically with modulation frequency, as expected.

ii. **Echo**

Table 9.5 shows an example of an MTF matrix computed from the same impulse response with an echo consisting of an impulse arriving 100 ms after an initial impulse. At low modulation frequencies the modulations of the initial and echo impulses are mostly in phase. At a modulation frequency of 10 Hz, the echo and the initial impulse are in phase (but 360 degrees displaced) and the modulation ratio is almost one. However, with a modulation frequency of 5 Hz, the echo is 180 degrees displaced and the signals mostly cancel, producing a low-modulation ratio compared with Table 9.4.

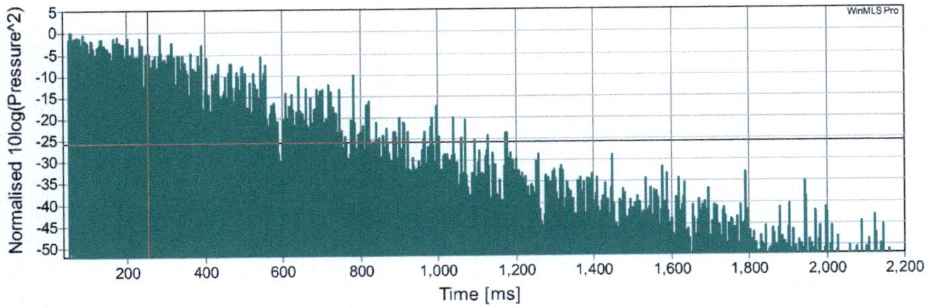

Fig. 9.13 Reverberated impulse in dB with an RT30 of 2.1 s

Table 9.4 Example of MTF matrix computed from a reverberated impulse with an RT30 of approximately 2.1s in each octave band

Modulation frequency	Octave band frequency (Hz)						
	125	250	500	1000	2000	4000	8000
0.63 Hz	0.913	0.899	0.895	0.906	0.907	0.91	0.903
0.80 Hz	0.876	0.856	0.851	0.866	0.867	0.871	0.86
1.00 Hz	0.828	0.8	0.795	0.815	0.818	0.82	0.806
1.25 Hz	0.766	0.725	0.727	0.752	0.759	0.758	0.74
1.60 Hz	0.676	0.617	0.635	0.668	0.684	0.675	0.652
2.00 Hz	0.574	0.496	0.539	0.583	0.615	0.592	0.567
2.50 Hz	0.457	0.368	0.436	0.495	0.551	0.508	0.484
3.15 Hz	0.333	0.287	0.349	0.41	0.488	0.425	0.41
4.00 Hz	0.277	0.292	0.28	0.327	0.409	0.348	0.346
5.00 Hz	**0.325**	**0.327**	**0.227**	**0.246**	**0.313**	**0.269**	**0.278**
6.30 Hz	0.286	0.337	0.239	0.169	0.227	0.196	0.176
8.00 Hz	0.28	0.328	0.291	0.2	0.206	0.227	0.196
10.00 Hz	**0.28**	**0.308**	**0.224**	**0.154**	**0.175**	**0.213**	**0.191**
12.50 Hz	0.118	0.19	0.16	0.07	0.15	0.136	0.116
MTI	0.51	0.5	0.49	0.48	0.51	0.5	0.49
STI	0.5						

Table 9.5 Example of MTF matrix with a single echo at 100 ms

Modulation frequency	Octave band frequency (Hz)						
	125	250	500	1000	2000	4000	8000
0.63 Hz	0.977	0.978	0.978	0.978	0.978	0.978	0.978
0.80 Hz	0.967	0.967	0.967	0.967	0.967	0.967	0.967
1.00 Hz	0.946	0.946	0.946	0.946	0.946	0.946	0.946
1.25 Hz	0.919	0.92	0.92	0.92	0.92	0.92	0.92
1.60 Hz	0.874	0.875	0.875	0.875	0.875	0.875	0.875
2.00 Hz	0.807	0.808	0.808	0.808	0.808	0.808	0.808
2.50 Hz	0.707	0.708	0.708	0.708	0.708	0.708	0.708
3.15 Hz	0.555	0.556	0.557	0.557	0.557	0.557	0.556
4.00 Hz	0.334	0.335	0.335	0.335	0.335	0.335	0.334
5.00 Hz	**0.162**	**0.163**	**0.163**	**0.163**	**0.163**	**0.163**	**0.16**
6.30 Hz	0.419	0.422	0.423	0.424	0.424	0.423	0.422
8.00 Hz	0.802	0.813	0.816	0.817	0.817	0.816	0.817
10.00 Hz	**0.966**	**0.986**	**0.992**	**0.993**	**0.994**	**0.993**	**0.994**
12.50 Hz	0.673	0.695	0.701	0.702	0.703	0.702	0.702
MTI	0.7	0.7	0.7	0.7	0.7	0.7	0.7
STI	0.7						

Prediction of STI with a Statistical Reverberant Field

The STI performance of a transmission channel can be predicted under the assumption that the reverberant field is a theoretically ideal diffuse field, with reverberation times as determined by statistical equations (see Chap. 2, Appendix 1).

Equation (9.3) shows the components of a simplified (scaled) impulse response for a room in which this assumption is valid.

$$h(t)^2 = I_D\delta(t) + I_R\tfrac{13.8}{T}e^{-13.8t/T}, \tag{9.3}$$

in which

- $\delta(t)$ is the Dirac delta function,
- I_D is the mean-square sound pressure due to direct sound at the receiver,
- I_R is the mean-square sound pressure due to reverberant sound,
- T is the reverberation time.

In practice, when using this method I_D is often associated with the level of direct sound plus the level of "early" reflections—typically those occurring within 50 ms of the direct sound—which serve to reinforce the direct sound.

Inserting Eq. (9.3) into Eq. (9.2) gives

$$m_k(f_m) = \tfrac{\sqrt{A^2+B^2}}{C}, \tag{9.4}$$

where

$$A = I_D + I_R\left[1 + \left(\frac{2\pi f_m T_k}{13.8}\right)^2\right]^{-1},$$

$$B = I_R\frac{2\pi f_m T_k}{13.8}\left[1 + \left(\frac{2\pi f_m T_k}{13.8}\right)^2\right]^{-1},$$

$$C = I_D + I_R + I_N.$$

I_N is the mean-square sound pressure due to noise and T_k is the reverberation time in octave band k. Although I_N could include the "noise" due to masking and the human hearing threshold, in common use I_N simply represents the octave-band level of acoustic noise present during the speech.

Note that this method is unable to consider the effects of late-arriving direct-field sound (for example from multiple loudspeakers) and echoes.

The spreadsheet Supplementary file1.xlsx calculates modulation transfer functions according to Eq. (9.4) and calculates STI from these.

References

Acoustical Society of America (2020) ASA/ANSI S3.5-1997 methods for calculation of the speech intelligibility index. Acoustical Society of America

Bies D, Hansen C (2009) Enginering noise control, 4th edn. Spon Press

Bild E, Redman A, Newman E, Muir B, Tait D, Schwarz N (2021) Sound and credibility in the virtual court: low audio quality leads to less favorable evaluations of witnesses and lower weighting of evidence. Am Pyschological Assoc 45(5):481–495

Houtgast T, Steeneken H (1985) A review of the MTF concept in room acoustics and its use for estimating speech intelligibility in auditoria. J Acoust Soc Am 77(3):1069–1077

International Electrotechnical Commission (2020) IEC 60268-16 2020. Sound system equipment—part 16: objective rating of speech intelligibility by speech transmission index. International Electrotechnical Commission

Krause J (2001) Properties of naturally produced clear speech at normal rates and implications for intelligibility enhancement. Ph.D. thesis, Massachusetts Institute of Technology

Morales K, Li F (2018) A new verification of the speech transmission index for the English language. Speech Commun 105:1–11

Standards Australia; Standards New Zealand (2016) AS/NZS 2107:2016 acoustics—recommended design sound levels and reverberation times for building interiors

Chapter 10
Acoustics and Sustainable Design

Abstract This Chapter describes acoustic issues related to the development of sustainable building practice. These include the use of natural ventilation, which can allow increased intrusive noise, natural lighting which can limit opportunities to provide acoustic absorption and lack of background sound from mechanical services which can make some spaces less comfortable. Potential tools and measures to deal with these issues are described.

10.1 Challenges to Traditional Acoustic Design

The traditional approach to acoustic design in buildings, as outlined in the preceding sections, has been:

- Control background sound, by design of mechanical services (largely air-conditioning systems) and by façade design to control external noise,
- Control intruding noise to an appropriate level (relative to the background) by designing internal partitions and external facades and
- Control sound quality by internal finishes and room design, with particular attention to speech intelligibility and/or music quality where required.

This approach assumes that the acoustic designer can nominate features of the ventilation and climate control system in the building, as well as the layout of spaces and their internal features, to achieve acoustic goals. In many contemporary buildings, this ability is limited due to the need to incorporate features of sustainable design, including:

- **Passive ventilation systems**. These involve allowing outside air to circulate in the building, which means outside noise will also enter unless this is controlled. It also means the removal of whole-of-building air-conditioning systems, which traditionally supply background sound to spaces where this is needed.
- **Chilled beam cooling systems**. These systems provide cooling with minimal or no air movement, and hence also remove a traditional source of background sound.

R. Bullen, *The Architecture of Sound*, Springer-AAS Acoustics Series,
https://doi.org/10.1007/978-981-97-9380-8_10

- **Natural lighting**, including large areas of glazing and skylights, as well as light-reflecting surfaces. Currently these features cannot be made acoustically absorptive, and hence they limit the possibilities for reverberation control.
- **Thermal mass**, typically exposed concrete or brick, for climate control. These materials cannot currently be made acoustically absorptive.

The above features are an integral part of a movement towards building design that is sustainable and consistent with environmental goals. Acoustic design and practice must adapt to this reality. This section describes some developing responses to the changing face of building design.

10.2 Passive Ventilation

Figure 10.1, adapted from (Acred and Hunt 2013), shows a typical system for passive ventilation in a multistorey building. Air enters each building level through vents, and leaves via an atrium connecting all levels.

Obvious acoustic issues arising from this design are:

- Entry of intruding sound from outside through the vents,
- Flanking transmission between building levels via the atrium,
- If the atrium area is used at the ground floor (which is common), intruding sound from noise in this area to other building levels and
- Lack of any controlled background sound for spaces that require it.

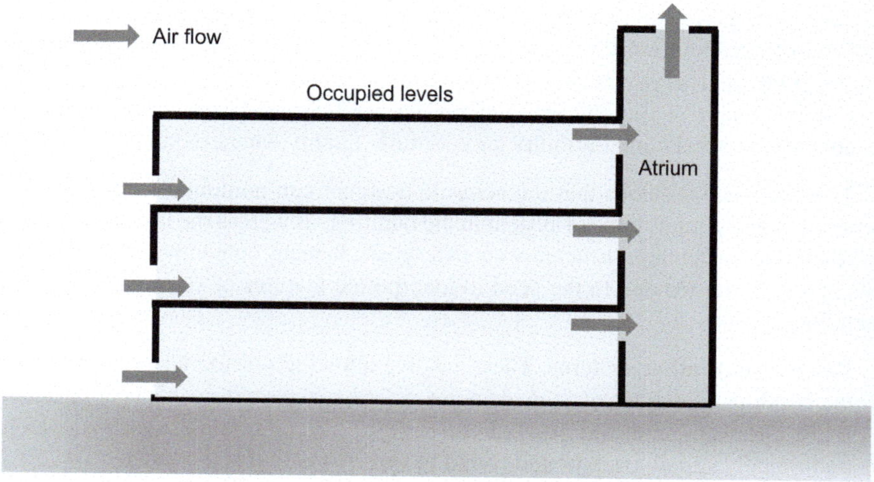

Fig. 10.1 Schematic of a typical large passively ventilated building (adapted from Acred and Hunt 2013)

The last point is discussed separately in Sect. 10.5. Regarding the other points, options for controlling intruding sound while allowing airflow include the following.

10.2.1 Acoustic Louvres

Standard adjustable louvres for control of airflow provide minimal sound reduction. However, if the blades are sufficiently long and incorporate acoustic absorption, they can provide some reduction (see Fig. 10.2). These "acoustic louvres" have traditionally been used to reduce noise emission from plant rooms and external plant items but can also be used to control sound transmission in buildings.

Louvres such as those shown Fig. 10.2 have a measured weighted sound reduction index of between R_w 12 and R_w 30 depending on the design. That is, the maximum reduction is still less than standard 6 mm laminated glazing. However, in some circumstances this may be all that is required.

These louvres add pressure drop to the ventilation system (with higher pressure drop at higher sound reduction) and hence may compromise the performance of the system. This needs to be assessed by a ventilation engineer.

Fig. 10.2 Acoustic louvres. The unit shown is 100 mm deep Image courtesy of https://airocle.com.au/

A further point is that, currently, acoustic louvres are not adjustable. A separate element is therefore required to adjust the airflow as part of a climate control system.

10.2.2 Acoustic Ventilators

These are typically smaller than louvres and are generally adjustable. An example is shown in Fig. 10.3. A large number of these units would be required to provide sufficient airflow for a building design as in Fig. 10.1. For the same airflow, their acoustic performance and pressure drop are comparable to acoustic louvres.

Duct silencers, as described in Sect. 4.5, could also be used to provide higher attenuation, but these would be bulky and generally result in unacceptable pressure drops.

Fig. 10.3 Acoustic ventilator. Image: Shutterstock

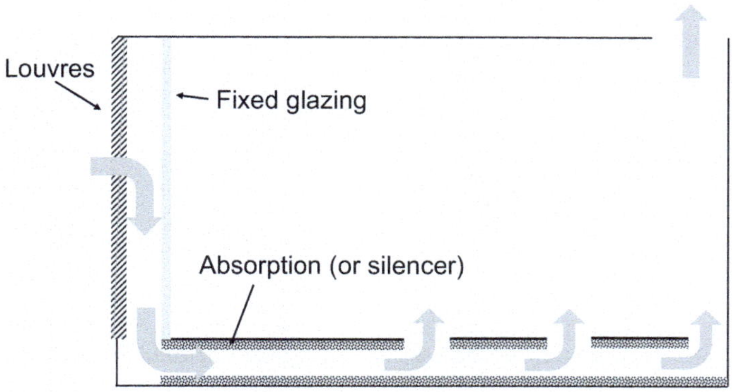

Fig. 10.4 Sound control with under-floor ventilation

10.2.3 Under-Floor Designs

For control of external noise entering the building, a popular design is shown in Fig. 10.4. Providing air from under the floor is often beneficial for the ventilation design, and also offers the chance to use lined ductwork or small silencers which can be designed as for an air-conditioning system (see Chap. 4) to reduce the level of intruding external noise. Of course, as well as reducing intruding noise, all these options also reduce the level of any constant background sound arriving from outside which may provide masking of internally generated sounds.

10.3 Chilled Beam Cooling Systems

This is not a new technology, but it is becoming more common, particularly in conjunction with passive ventilation systems, in an effort to further reduce energy usage. The two main systems are illustrated in Fig. 10.5. Passive chilled beam systems do not require a primary air fan, but are generally limited to small spaces. Active systems use a primary air fan but do away with VAV boxes in the ceiling space (see Sect. 4.11).

Acoustically, the main impact of these systems is to remove a traditional source of background sound. Consequences of this are discussed in Sect. 10.5.

10.4 Natural Lighting and Thermal Mass

Making as much use as possible of natural lighting in a building means introducing large areas of glazing and light-reflecting surfaces. Figure 10.6 shows an example.

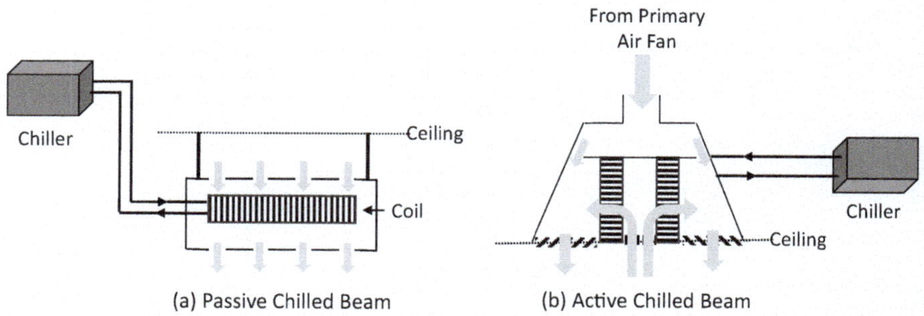

Fig. 10.5 Chilled beam cooling systems—**a** passive; **b** active (adapted from ASHRAE, ASHRAE Handbook 2019)

Fig. 10.6 Use of natural lighting Image: Shutterstock

Here it is obvious that opportunities for locating acoustic absorption are limited, and without acoustic treatment there will be significant reverberation, with consequent reduction in speech intelligibility.

The use of large areas of exposed mass—generally concrete—helps to stabilise the thermal environment, once again resulting in lower energy use. Figure 10.7 shows an example. This technique further reduces the opportunities for incorporating acoustic absorption.

In these cases, where low reverberation is important for the functioning of the space there is no option but to include high-grade absorbing material in whatever space is available. Office ceilings are an obvious choice in Fig. 10.6, but this

Fig. 10.7 Exposed mass ceiling and walls Image: Shutterstock

still leaves multiple, potentially parallel glass surfaces, and this is not an option in Fig. 10.7. Any surface that is not required to be light-reflecting or include exposed mass needs to be targeted for absorption.

Microperforated absorbers (see Chap. 8 Appendix 1) provide the possibility of achieving reasonable levels of sound absorption in a light-reflective surface.

Where this is not sufficient, absorbing materials need to be provided as part of the room contents. A popular option is hangings (e.g. Fig. 10.8) of which there are numerous commercially available alternatives.

The interface between energy-saving building design and acoustics is still being investigated. It is likely that new products and/or materials will be required to produce buildings that have low or zero carbon footprint but support a comfortable acoustic environment.

10.5 Background Sound and Artificial Masking

Removal of building-scale air-conditioning systems means removal of the most important traditional source of background sound in large buildings. In many cases this makes achievement of the minimum levels of background sound recommended in AS/NZS 2107, as described in Chap. 5, difficult.

It may be questioned whether achievement of these minimum values is really required. They represent the minimum background levels typically found in the relevant spaces, and where:

- intruding sound can be sufficiently controlled,

Fig. 10.8 Absorptive panels hung from an exposed ceiling Image courtesy of Envirotech Acoustics

- there will be little or no distracting noise created within the space and
- speech privacy within the space is not important.

there appears to be no good reason why lower background sound levels should not be acceptable, or even appreciated. (A conference room would be an example of such a space.) Further, there are suggestions that occupants of naturally ventilated buildings may have a different tolerance to intrusive noise compared with those in traditional spaces (Field 2015).

Nevertheless, there is substantial evidence—for example, (Hongisto 2008)—that in environments where many people share the same space but may be engaged in different activities, background sound makes the space more usable and more pleasant. Such spaces include open-plan offices, some educational spaces and hospital wards.

In such cases, increasing use is being made of artificial masking sound. This will be presented by loudspeakers, generally located in the ceiling space.

The sound will be constant or almost constant, producing a spectrum in the occupied space that is designed to provide maximal masking of typical sounds (usually speech) with minimal distraction. Sound levels are generally set at between 40 and 45 dBA. It is generally considered that masking sound greater than 45 dBA will be a source of annoyance and is likely to cause people to speak more loudly, negating the masking effect.

Suppliers of masking systems all have tweaks in their recommended spectra and installation procedures that are claimed to provide better masking. However, the spectrum in Fig. 10.9, taken from (Taylor 2000), is typical. It is similar to the spectrum

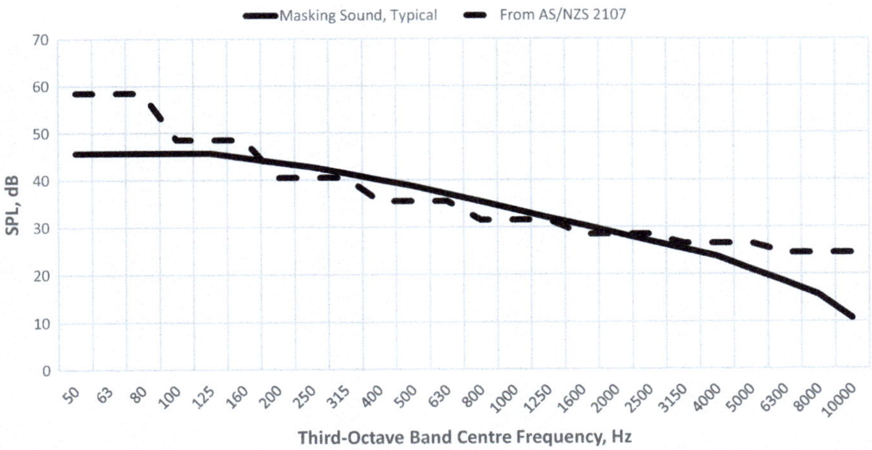

Fig. 10.9 Typical spectrum of masking sound (from Taylor 2000) and recommended background spectrum shape from AS/NZS 2107 Appendix C, both at 45 dBA

recommended in AS/NZS 2107:2016 (Standards Australia 2016) (converted to third-octave bands), also shown in Figure 10.9, which is oriented to controlling sound produced by an air-conditioning system, except at very low and very high frequencies.

Of course, the spectrum in Fig. 10.9 is not the spectrum emitted by the loudspeakers. Allowance must be made for the sound reduction index of the ceiling at each frequency and the acoustic properties of the space so that the desired spectrum is broadcast into the space.

10.6 Summary

The apocryphal curse "May you live in interesting times" applies to current building acoustics. As the principles of sustainable building design are (thankfully) being implemented in an increasing number of building projects, including renovations as well as new builds, new ways to provide a high-quality acoustic environment in these buildings need to be investigated.

The tools required are a combination of existing and new or newly popular elements including:

- acoustic louvres and ventilators used in unusual ways,
- high-quality absorbers in unusual places,
- alternative materials such as microperforated absorbers,
- sound masking and
- probably other tools not yet considered.

The important point is that each space requires its own acoustic environment, which is a combination of background sound, intruding sound and wanted sound, together with indicators of sound quality. The designer needs to first understand the acoustic requirements of each space in terms of these soundscape elements, and then select appropriate tools to achieve them, compatible with the other constraints on the design.

References

Acred A, Hunt G (2013) Passive ventilation in multi-storey atrium buildings: a first-order design guide. In: Proceedings of 34th AIVC conference

ASHRAE, ASHRAE Handbook—HVAC applications chapter 49: noise and vibration (2019)

Field C (2015) Acoustic design criteria for naturally ventilated buildings—unnecessarily stringent?. [Online]. Available: airahfiles.org.au/Ecolibrium/2015/02-15-Eco-004.pdf

Hongisto V (2008) Effects of sound masking on workers—a case study in a landscaped office. In: 9th international congress on noise as a public health problem

Standards Australia (2016) Standards New Zealand, AS/NZS 2107:2016 acoustics—recommended design sound levels and reverberation times for building interiors

Taylor A (2000) Sound masking systems [Online]. Available: www.atlasied.com/d/3781/ATS005 388A-Tech-Guide-for-Effective-Sound-Masking.pdf

Part III
Design in Specific Spaces

Chapter 11
Classrooms

Abstract In this Chapter, the acoustics of classrooms is considered in depth, using as a guide the principles outlined in Part 2. Criteria and recommendations for background sound levels, intruding sound, reverberation and speech intelligibility are discussed and compared with measured values reported in real classrooms. Research relating to the impact of classroom acoustics on teachers and students is described, and best practice for acoustic design is summarised.

11.1 Approach to Acoustic Design

Teaching spaces include (among others).

- classrooms (enclosed and open-plan; primary and secondary),
- gyms and sports halls,
- music rooms,
- pre-school play areas,
- lecture theatres.

These spaces all have specific, and different, acoustic requirements. This chapter will concentrate on classrooms, which are well-studied and have the most unique acoustic requirements.

Desired outcomes for any space should always be determined by those with responsibility for its use—in this case, teachers. However, in the apparent absence of any formal study of teachers' preferences regarding the acoustics of classrooms, the following can be assumed to be generally agreed goals.

- Speech intelligibility: There should be minimal loss of words spoken by the teacher when heard anywhere in the classroom, including for students from non-English speaking backgrounds.
- Comfort and attention span: Acoustic conditions in the classroom should minimise distraction and disturbance, and promote comfort and focus.

R. Bullen, *The Architecture of Sound*, Springer-AAS Acoustics Series,
https://doi.org/10.1007/978-981-97-9380-8_11

- Stress for teachers: Teachers should not be required to raise their voice for prolonged periods, should not be distracted by other sounds, and should hear students clearly when they speak.

As for any space, acoustic design can proceed by considering the major acoustic elements.

Background Sound: (that is, constant sound when the space is unoccupied).

- How loud should it be?
- Are there important characteristics apart from its level?
- What are the consequences if the background sound is too loud or too quiet?
- How can it be controlled?

Intruding Sound: (i.e. sound from the environment, from other rooms or from other activities in the same room).

- How loud can it be without disruption?
- What are the consequences if it is too loud?
- How can it be controlled?

Sound Quality:

- What is an appropriate RT?
- What is an appropriate STI?
- Are there other quality-related factors that are also important?
- What are the consequences if sound quality is inappropriate?
- How can it be controlled?

The following sections will describe guidance on these questions with respect to the specific needs of classrooms.

11.2 Background Sound in Classrooms

11.2.1 Recommended Background Sound Levels

Table 11.1 shows guidance from various documents as to appropriate sound levels in classrooms. Other documents have recommendations that range from 30 to 45 dB in various acoustical units. No standard or recommendation appears to include criteria other than SPL values in dBA.

Table 11.1 Recommended background sound levels in classrooms

Document	Description of space	Recommended background SPL
AS/NZS 2107:2016 (Standards Australia 2016)	Teaching spaces/single classroom: open plan teaching spaces; primary schools; secondary schools	Range 35–45 dBA L_{Aeq}
AAAC Guideline for Educational Facilities (Association of Australasian Acoustical Consultants)	Open plan	$\leq$40 dBA L_{Aeq}
	Primary schools; secondary schools	35–40 dBA L_{Aeq}
	Hearing impaired	$\leq$ 30 dBA L_{Aeq}
U.K. building bulletin 93 (U.K. Department of Education 2015)	Classroom, new build	$\leq$ 35 dBA $L_{Aeq,30\ min}$
	Classroom, refurbishment	$\leq$ 40 dBA $L_{Aeq,30\ min}$
	Open plan, new build	$\leq$ 40 dBA $L_{Aeq,30\ min}$
	Open plan, refurbishment	$\leq$ 45 dBA $L_{Aeq,30\ min}$
USA–ANSI S12.60–2010 (ANSI)	Core learning space	$\leq$ 35 dBA $L_{Aeq,1\ h}$

11.2.2 Measured Background Sound Levels

Table 11.2 shows reported measurements of background sound levels in classrooms. From these results it can be concluded that background sound levels in the majority of classrooms are within or below the range 35–45 dBA L_{Aeq} as recommended in AS/NZS 2107:2016, but some exceed this range, sometimes by up to 10 dB (Figs. 11.1 and 11.2).

11.2.3 Impacts of Background Sound

One study (Shield et al. 2015) reports a small but significant correlation between background sound levels and sound levels during lessons. These results are shown in Fig. 11.3.

Of course, the measured background sound levels were not directly influencing the noise levels during the lessons, which are at least 10dB higher. The authors' suggestion is that higher background noise levels are associated with more unruly behaviour or "disruptive activities" which is evidenced in a higher noise level in the classroom. (They also found a small but significant correlation between background noise level and "total disruption time" in a lesson.) Potentially this could be associated with lower intelligibility of speech from the teacher, which would be expected at higher background sound levels (Chap. 10).

Table 11.2 Measured background sound levels in classrooms

Location and reference	Conditions	Measured background SPL
London (Shield and Dockrell 2004)	Average of 30 classrooms in 16 primary schools	47 dBA L_{Aeq}*
Ottowa, Canada (Sato and Bradley 2008)	Average of 16 primary school classrooms	42 dBA L_{Aeq}
England (Shield et al. 2015)	Average of 13 secondary schools	Enclosed classrooms—34 dBA L_{Aeq}, range 27–42 dBA
		Open plan classrooms—35 dBA L_{Aeq}, range 31–45 dBA
Ohio, U.S.A. (Knect et al. 2002)	Levels as shown in Fig. 12.1	
Brisbane, Australia (Wilson et al. 2019)	Levels as shown in Fig. 12.2	
Sydney, Australia (Mealings et al. 2015a)	Four classrooms	36–46 dBA (presumed L_{Aeq}) highest level "most likely" due to HVAC

* Note the average L_{A90} was 37 dBA, so the noise was not steady and potentially not true background sound

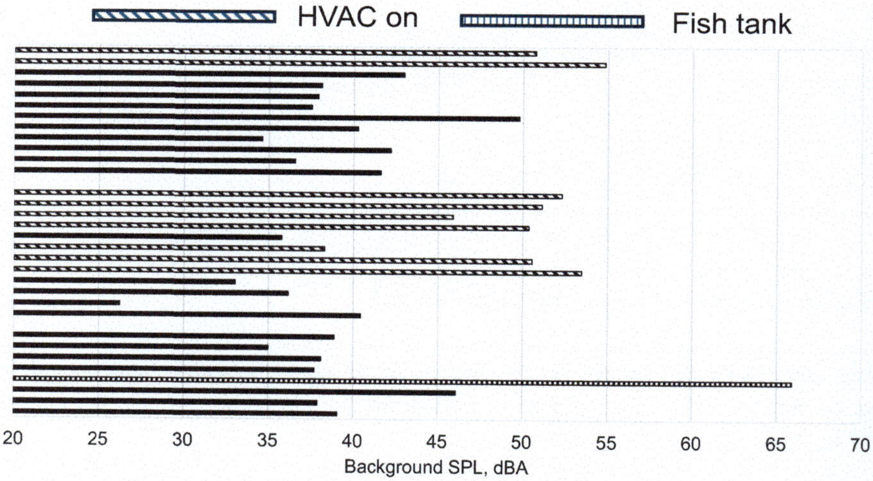

Fig. 11.1 Measured background sound levels in unoccupied classrooms, adapted from Knect et al. (2002)

From Fig. 11.3, background noise is clearly only one of many factors influencing noise levels during a class. However there is a suggestion that it may have an effect on classroom dynamics at levels greater than about 40–45 dBA L_{Aeq}. Another study (Mealings et al. 2015a) with background levels all below 40 dBA showed no relationship with sound levels during a class.

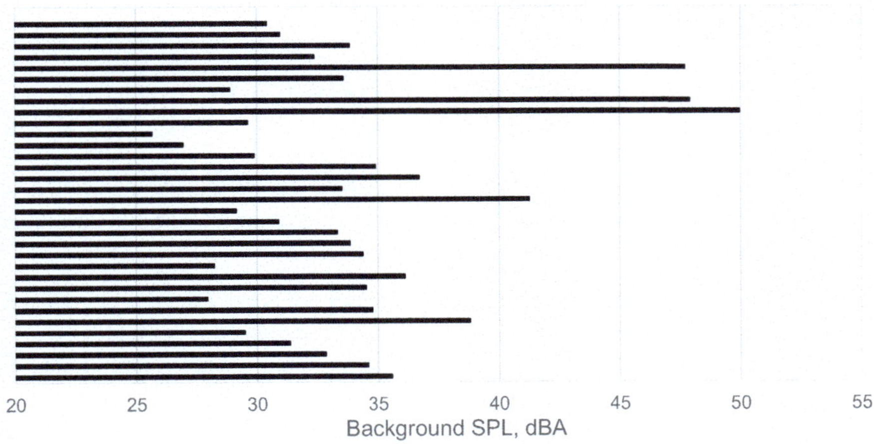

Fig. 11.2 Measured background sound levels in unoccupied classrooms, adapted from Valentine and Wilson (2002)

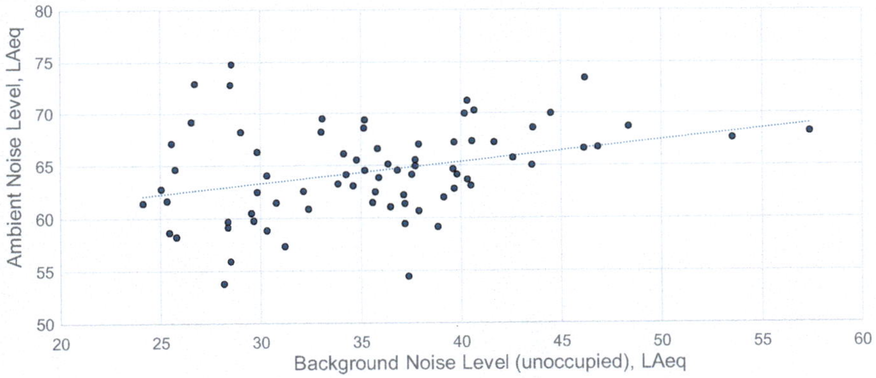

Fig. 11.3 Relationship between background noise level (in unoccupied classroom) and ambient noise level during lesson, adapted from Shield et al. (2015)

Where background noise is above 45 dBA this appears to be generally due to HVAC systems. Control in this case is a matter of ensuring that building contracts include acoustic standards equivalent to AS/NZS 2107 and that these standards are enforced. There is a suggestion from the above data and from speech intelligibility studies described in Chap. 10 that 40 dBA may be a more appropriate upper limit than 45 dBA for these spaces.

A number of companies offer sound masking systems for teaching spaces, which would increase background levels where they are low—generally to about 40 dBA. The companies offer anecdotal evidence of their effectiveness in masking distracting noise. However there appear to be no independent studies of the effectiveness of these systems in teaching environments. Their use is currently rare, at least in Australia.

11.3 Intruding Noise—Environmental

Figure 11.4, from a New Zealand study (Valentine and Wilson, 2002), shows teachers' reports of the sources of intrusive external noise in the classroom. These will obviously differ depending on the school's environment.

One source of intruding noise whose effects have been extensively studied is transportation noise. (Stansfeld et al. 2005) reports results from a very large European study, the major findings of which can be summarised as:

- External aircraft noise above 55 dBA L_{Aeq} at a school is associated with reduced reading scores in children (after controlling for numerous other variables).
- Some effects of aircraft noise were found in memory and other tests but these are not consistent.
- No consistent effect was found for road traffic noise.

The effect of aircraft noise on reading scores is shown in Fig. 11.5. Note that 55 dBA L_{Aeq} is approximately 20 ANEF which is the "acceptable" limit for new schools under Australian Standard 2021. However there are numerous existing schools located in areas with aircraft noise above this level.

Whether this effect is mitigated by reducing internal aircraft noise levels in classrooms is an open question—it may be due at least in part to noise exposure in other places. The design of measures to control internal aircraft noise levels is described in AS 2021.

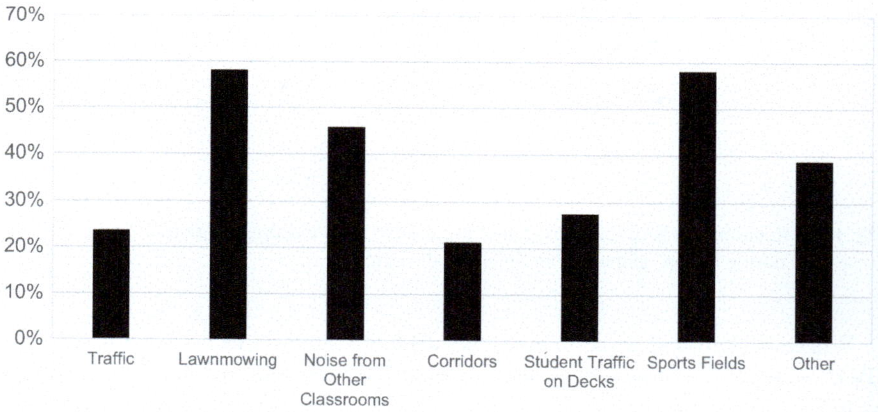

Fig. 11.4 Percentage of concerned teachers reporting various sources of intruding noise in classrooms, adapted from (Valentine and Wilson, 2002)

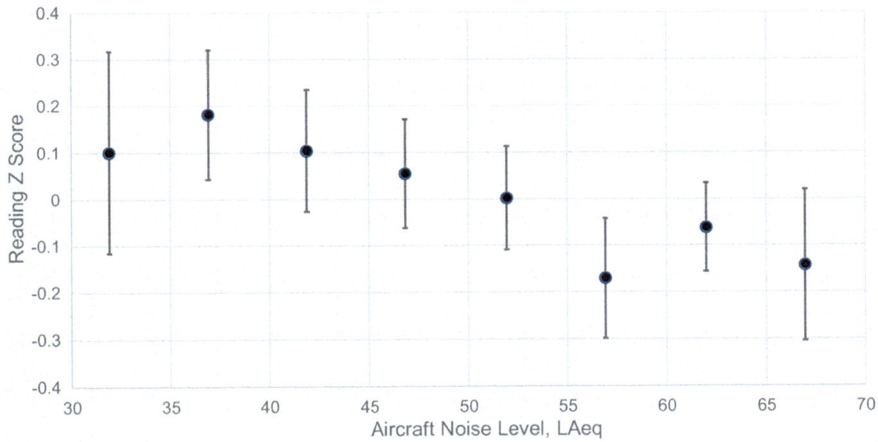

Fig. 11.5 Adjusted mean and 95% confidence intervals for reading scores compared with aircraft noise L_{Aeq} at schools, adapted from Stansfeld et al. (2005)

11.4 Intruding Noise from Other Classrooms

Noise intrusion from other classrooms can be assessed using the techniques described in Chap. 7. Measurements reported in Table 11.3 indicate that sound levels in a noisy classroom can be 77 dBA L_{Aeq}. Hence L_{Amax} values could be estimated at 80–85 dBA.

Assuming a background level of 35 dBA and the "rule of thumb" that intrusive noise should not exceed background noise by more than 5 dBA, this implies that $D_{nT,w} + C$ of the partition between classrooms should be at least

$$D_{nT,w} + C = 85 - (35 + 5) = 45.$$

This is consistent with other recommendations:

AAAC (Association of Australasian Acoustical Consultants): $D_w \geq 45$,
UK Building Bulletin 93 (U.K. Department of Education 2015): $D_{nT,w} \geq 45$,
ANSI S12.60–2010 (ANSI): STC > 50 (approximately $D_{nT,w} \geq 45$).

Table 11.3 Average recorded noise levels in classrooms in London schools (Shield and Dockrell 2004)

Activity	L_{Aeq}, dBA
Silent reading/test	56
1 person speaking	61
Individual work	65
Individual work and movement	72
Group work	73
Group work and movement	77

Constructions achieving $D_{nT,w}$ 40–45 include:

13 m plasterboard each side of 64 mm studs + insulation,

100 mm blocks rendered 13 mm on both sides.

Constructions achieving $D_{nT,w} > 45$ include:

2 × 13 m plasterboard each side of 64 mm studs + insulation,

230 mm blocks rendered 13 mm on both sides.

In both cases these $D_{nT,w}$ values assume that there are *no flanking transmission paths*.

A particular problem in classrooms is the use of operable walls, which require careful installation and good maintenance, and may be subject to various forms of flanking transmission. Well-designed operable walls such as Fig. 11.6 can achieve R_w values of 41–53. However data in Mealings et al. 2015a imply that the actual performance of an installed wall in a classroom was about $D_{nT,w}$25, due entirely to flanking transmission.

Fig. 11.6 Commercial sound-rated operable wall

11.5 Intruding Noise from the Same Room

Noise levels within an enclosed classroom obviously depend on the activities being undertaken. For example, Table 11.3, from data in Shield and Dockrell (2004), shows average recorded noise levels during various activities.

In open plan classrooms, sound from other classes in the same space effectively constitutes intruding noise. Mealings et al. (2015a) investigate sound levels in four classrooms:

- an enclosed classroom with 25 students,
- a double classroom with 4 students,
- a linear fully open-plan triple classroom with 91 students,
- a semi-open plan K-6 classroom with 205 students.

The study computed the "signal to noise ratio" (SNR)—the difference between the L_{Aeq} of the teacher's voice and the L_{Aeq} of sound from other classes—for various activities in the adjacent class. Results are shown in Fig. 11.7.

The ratios decline with increasingly noisy activity in the adjacent class, even in the enclosed classroom, indicating inadequate sound reduction performance of the partition. However, for the open-plan classrooms the decline is much steeper, with signal-to-noise ratios being negative in some cases. These results assume:

- there is no noise from the class itself;
- the teacher's voice level is constant for all scenarios.

In fact, teachers do raise their voice to compensate for intrusive noise as demonstrated in Fig. 11.8. This is an example of the "Lombard effect" under which people tend to speak more loudly when ambient noise levels are higher (Whitlock 2012). (They also alter other characteristics of their speech including a shift to higher

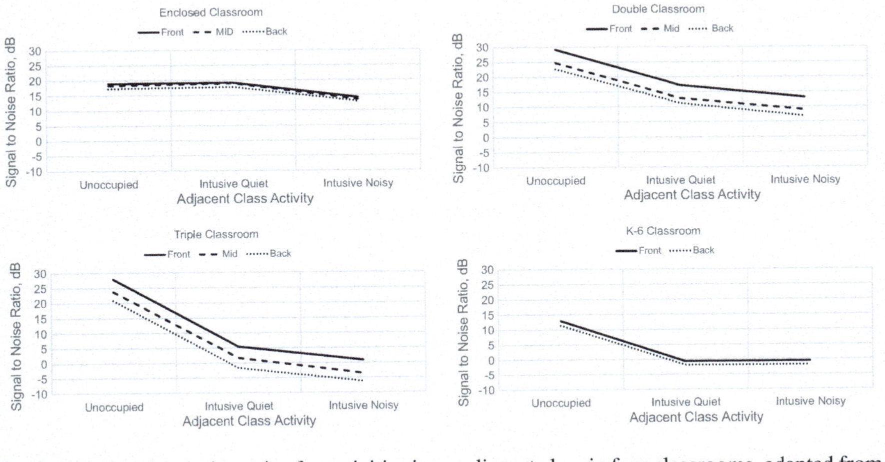

Fig. 11.7 Signal-to-noise ratios for activities in an adjacent class in four classrooms, adapted from Mealings et al. (2015a)

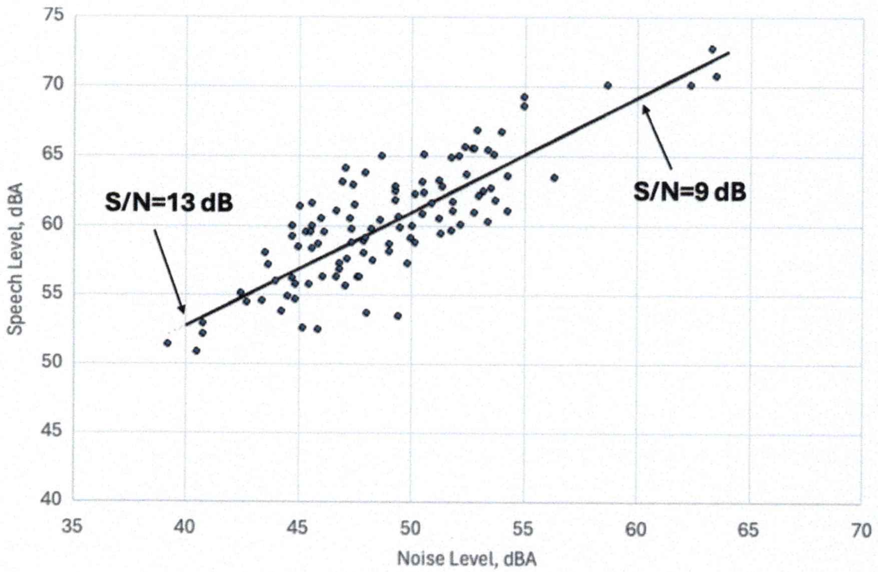

Fig. 11.8 Mean speech level of the teacher and mean noise level in 28 classrooms, adapted from Sato and Bradley (2008)

frequencies and an increase in the duration of vowels.) The same effect occurs for students, who speak more loudly, thus increasing the noise level further.

This of course adds further strain for the teacher. Figure 11.8 shows that the resulting increase in the teacher's voice level is not sufficient to maintain a constant SNR.

11.6 Effect of Noise on Speech Perception

(Mealings et al. 2015b) gives results of speech perception tests conducted in the four schools described above. The tests involved a variant of the word recognition tests described in Sect. 10.6.4. During these tests the noise level due to activity in adjacent classes was assessed at one point in each room. The sound level from speech was constant (from a loudspeaker) although distance from the speaker varied.

As expected, there were clear effects of both noise level and distance from the loudspeaker on perception scores, with perception score falling by 14% for each 10 dB increase in noise level and by 13% for each doubling of distance from the source. With an intruding noise level of 60 dBA and a distance of 2 m from the speaker, the perception score was 65% in the enclosed, double and triple classrooms, which would not generally be considered acceptable.

The speech perception scores of individual students were related to measured STI values as shown in Fig. 11.9, where the sigmoidal function represents average scores.

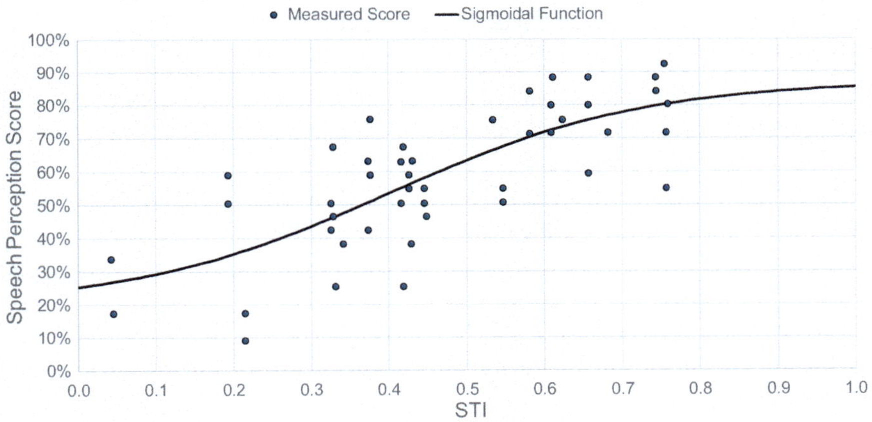

Fig. 11.9 Relationship between speech perception score and STI, adapted from Mealings et al. (2015b)

It is interesting to compare these scores with scores for adults on a standard PB word test, as shown Sect. 10.6.4. This comparison is shown in Fig. 11.10. Although the testing procedures differ, there is some evidence that children's scores tend to be lower at the same STI.

In the large K-6 classroom, scores were 8% higher for the same noise and distance values. The authors comment that "The most likely explanation for these better scores is that the design of the K-6 classroom aided the children's speech perception—this classroom was newly purpose-built as a 'twenty-first century learning space'".

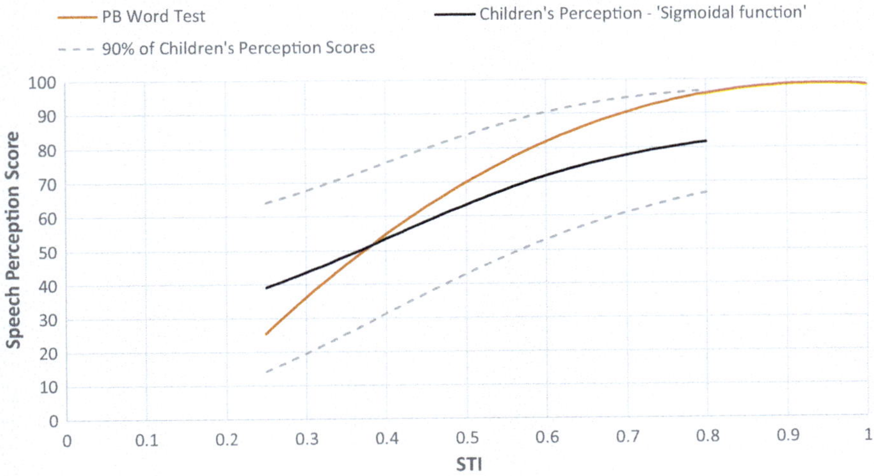

Fig. 11.10 Comparison of speech perception scores from children's testing (Mealings et al. 2015b) and standard PB word recognition (Morales and Li 2018)

11.7 Effect of Noise on Task Performance

Apart from their ability to hear and understand the teacher, the ability of students to perform tasks is also potentially affected by intruding noise. Here the evidence is mixed and somewhat confusing.

(Dockrell and Shield 2006) describes a study in which tasks were presented to groups of primary school children under three noise conditions:

- "Base"—normal classroom with everyone quiet (not measured but presumably approximately 50 dBA L_{Aeq}),
- "Babble"—unintelligible children's voices at 65 dBA L_{Aeq},
- "Babble + environmental"—as for "Babble" with additional random sirens and lorries at 58 dBA L_{Amax}.

Relevant tasks included reading and spelling. In addition to typical-needs children, there was a group of children with special educational needs.

Some results are shown in Fig. 11.11.

Task performance in the presence of "babble" decreased compared with the "base" case, particularly for those with special educational needs. However, the introduction of environmental noise in addition to "babble" appears to compensate, or more than compensate, for this reduction.

The meaning of this result is not clear, and further research is clearly required. For the present the message seems to be that impacts on task performance depend not so much on the level of intruding noise but more on the content of the noise.

11.8 Reverberation Times in Classrooms—Recommended

Reverberation times for classrooms are usually specified as "mid-frequency" or an average over 500 Hz–2 kHz. In almost all cases measurements are performed using RT_{30}, so in non-Sabine spaces (which would include most classrooms) "reverberation time" can be taken to mean RT_{30}.

Table 11.4 shows guidance as to appropriate reverberation times from several documents.

AS 2107 recommends the values shown in Fig. 11.12 for "Teaching/ Communications" and includes a note recommending that "certain teaching spaces, including those intended for students with learning difficulties and students with English as a second language, should have reverberation times at the lower end of the range". The German Standard DIN 18041 explicitly includes recommendations for "Teaching/Communications inclusive" as shown in Fig. 11.12, which are intended for similar types of teaching environment.

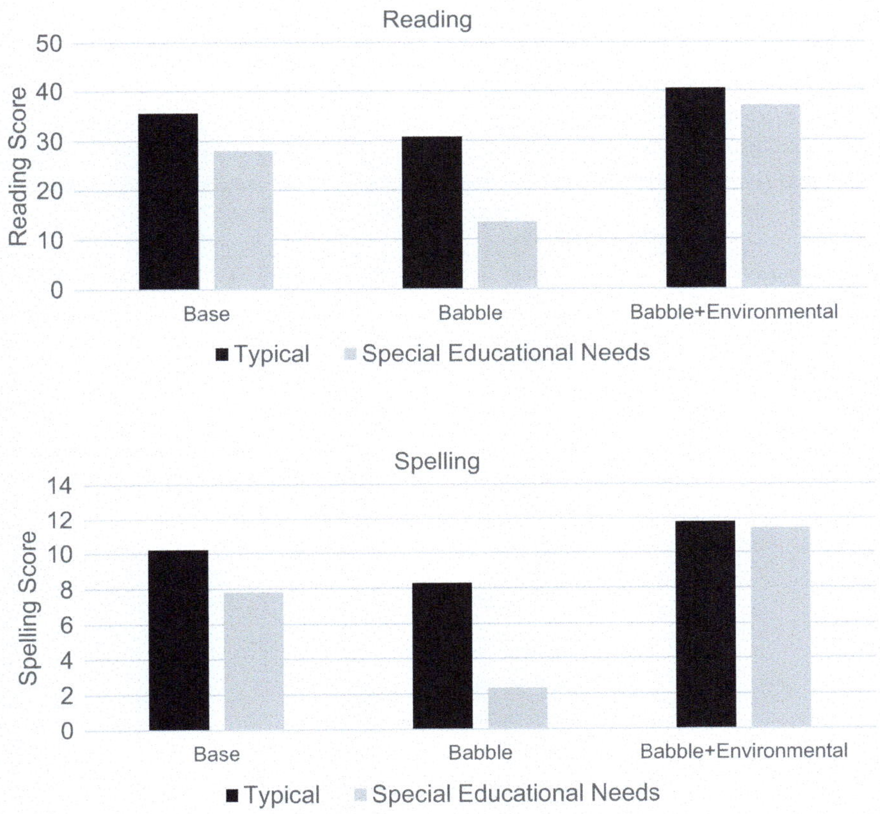

Fig. 11.11 Children's task performance under different noise conditions, adapted from Dockrell and Shield (2006)

11.9 Reverberation Times in Classrooms—Measured

Figure 11.13 shows measured reverberation times in classrooms in Ohio, U.S.A (Knect et al. 2002).

Figure 11.14 shows reverberation times measured in the four Sydney classrooms described in (Mealings et al. 2015a).

These and other references suggest that reverberation times typically vary from about 0.5 s to 1 s and are very often higher than values recommended in standards.

Table 11.4 Recommended reverberation times in classrooms

Document	Description of Space	Recommended Reverberation Time
AS/NZS 2107:2016 (Standards Australia 2016)	Teaching spaces/single classroom	Reverberation times shown in Fig. 11.12 (see text)
DIN 18041 (European Standards 2016)	Teaching/communication and teaching/communication inclusive	Reverberation times shown in Fig. 11.12 (see text)
AAAC guideline for educational facilities (Association of Australasian acoustical consultants, guideline for educational facilities; Association of Australasian acoustical consultants)	Open plan; primary schools; secondary schools	≤ 0.6 s
	Hearing impaired	≤ 0.4 s
U.K. Building Bulletin 93 (U.K. Department of Education 2015)	Primary classroom, new build	≤ 0.6 s
	Primary classroom, refurbishment	≤ 0.8 s
	Secondary classroom, new build	≤ 0.8 s
	Secondary classroom, refurbishment	≤ 1.0 s
	Open plan, new build	≤ 0.5 s
	Open plan, refurbishment	≤ 0.5 s
USA–ANSI S12.60–2010 (ANSI)	Core learning space < 283 m^3	≤ 0.6 s
	Core learning space > 283 m^3 and < 566 m^3	≤ 0.7 s
	Core learning space > 566 m^3	Absorbing materials recommended

11.10 Impact of Reverberation in Classrooms

(Tiesler et al. 2018) describes a study conducted in a German elementary school in which one classroom was retrofitted with acoustic treatment, reducing the reverberation time from 0.7s to 0.4s. Figure 11.15 shows measured noise levels during five consecutive teaching sessions under both conditions, beginning with the first session in the morning. (Note these are L_{A95} levels, not L_{Aeq}.)

In the first session, an increase in RT was associated with a small increase in noise level—roughly what would be expected if the sound power in the room were constant. However, as the day wore on, in the higher reverberation environment the class became noisier, while in the lower reverberation environment the level stayed approximately constant. This is highlighted in Fig. 11.16 which records the number of "dysfunctional activities" recorded during the same teaching sessions.

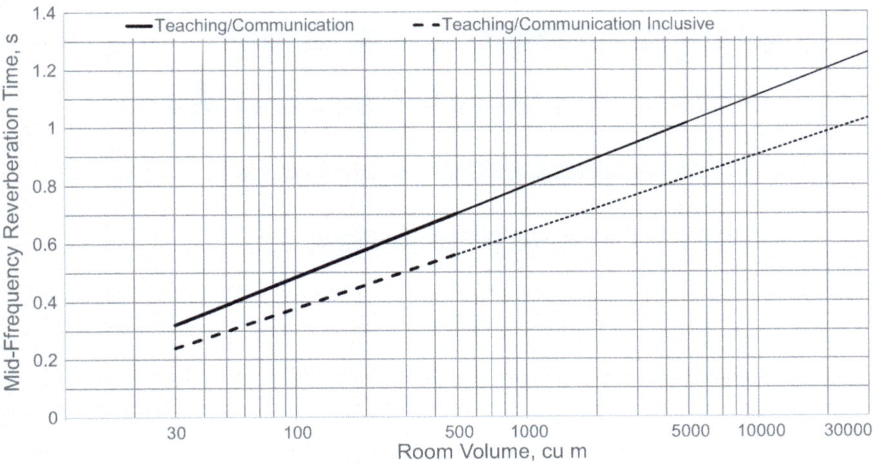

Fig. 11.12 Recommended reverberation times for: "teaching/communication" from AS/NZS 2107:2016 (Standards Australia 2016); and "teaching/communication inclusive" from DIN 18041 (European Standards 2016)

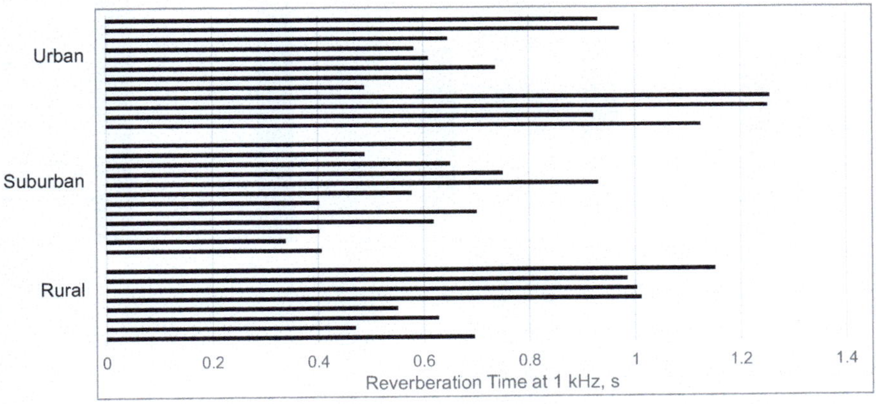

Fig. 11.13 Measured reverberation times in classrooms, adapted from Knect et al. (2002)

Even more strikingly, the study also recorded the teacher's heart rate at 5 min intervals, with some results shown in Fig. 11.17. In the lower-reverberation environment a 10 dB increase in noise level was associated with an increase of 4 beats/min in heart rate. In the higher-reverberation environment the same increase in noise level produced an increase of 10 beats/min in heart rate.

It seems clear that reverberation, together with perhaps other indicators of sound quality, directly affects both the functioning of a class and the stress levels of teachers. These effects go well beyond the direct effect of RT on sound level.

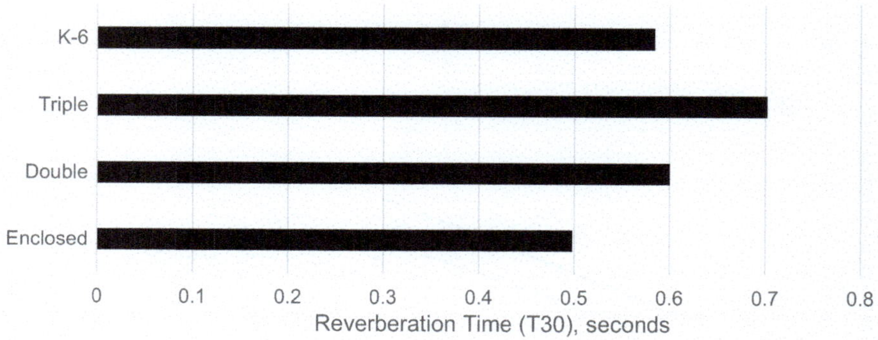

Fig. 11.14 Average measured reverberation times in classrooms, adapted from Mealings et al. (2015a)

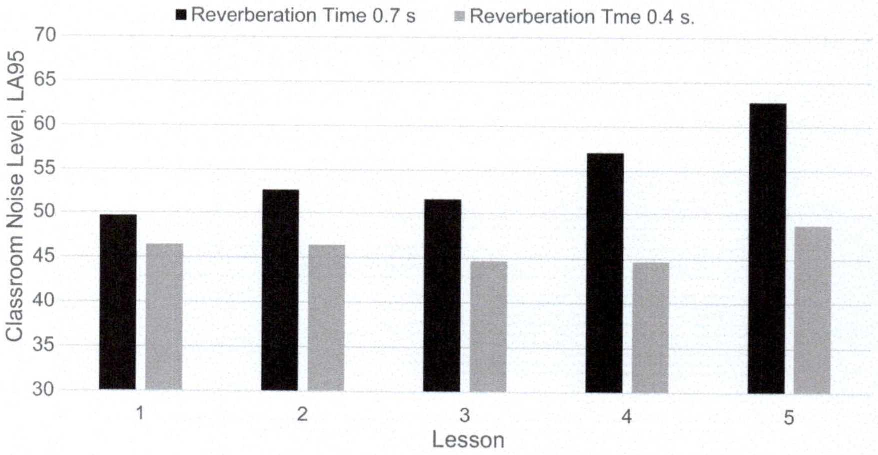

Fig. 11.15 Classroom noise levels during five consecutive lessons in higher and lower RT environments, adapted from Tiesler et al. (2018)

11.11 Other Measures of Sound Quality

Some researchers have considered other measures of sound quality in a classroom situation, particularly clarity (C_{50} as described in Sect. 8.3.2) (Christensson 2018) (Harvie-Clark 2014). As described in Chapters 8 and 9, this is likely to be linked not only with speech intelligibility but also with a sense of "presence" of the speaker.

There is a suggestion that C_{50} may provide a better indicator of "good acoustics" in a classroom than reverberation time (in any of its variants), but while plausible, there is as yet no firm evidence for this. In addition, there are no officially sanctioned design guidelines in terms of C_{50}. Some researchers (Harvie-Clark 2014) have suggested $C_{50} = 3$ dB as a minimum standard for classrooms. However, consideration of the

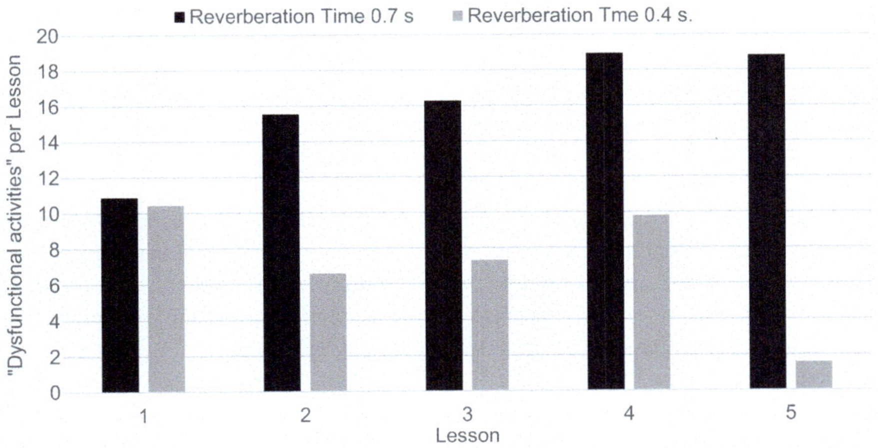

Fig. 11.16 "Dysfunctional activities" during five consecutive lessons in higher and lower RT environments, from Tiesler et al. (2018)

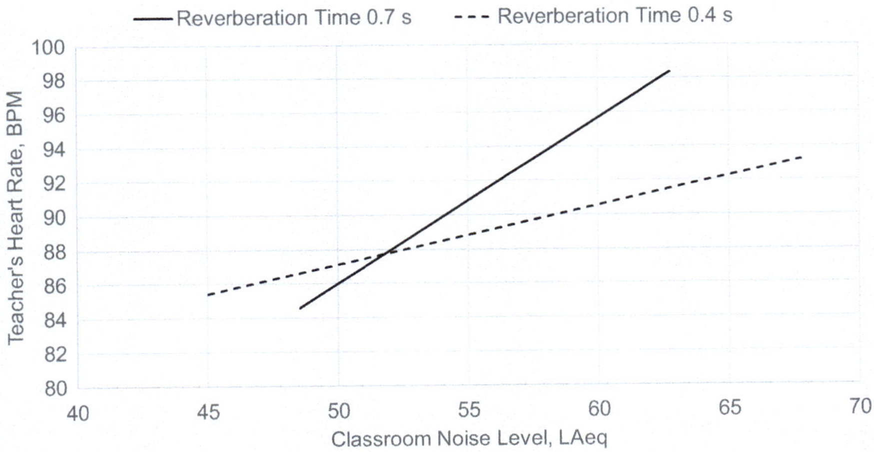

Fig. 11.17 Teacher's heart rate as a function of classroom SPL, from Tiesler et al. (2018)

simulated sounds in Table 2.1 of Chap. 2, Appendix 1 would suggest that $C_{50} = 6$ dB may be a more appropriate minimum, with $C_{50} = 9$ dB being a goal.

C_{50} should not be considered in isolation, as it does not take account of loudness, or gain (see Sect. 8.3.2), and hence a goal in terms of C_{50} does not control potential masking by other sound.

11.12 "Soundfield" Systems

One potential measure to increase STI and SNR is to give the teacher a microphone. These are known as "soundfield" systems. They can increase the teacher's voice level for the entire class or can allow the speech to be directed to a (selectable) small group only.

These systems are reasonably common in the U.S., but not in Australia except for hearing-impaired students. For children with a hearing aid or cochlear implant, the teacher's voice is transmitted directly by Bluetooth or WiFi. This has been reported to be very successful, with the proviso that the reception level, and potentially the frequency response, must be set individually for each student.

For normal-hearing students, there is insufficient data to provide a good evaluation. However the following comments are offered as suggestions.

For enclosed classrooms, the systems could provide better speech reception in the presence of intruding noise and/or high RT, and would remove the need for increased vocal effort by the teacher. Whether classroom noise levels would simply increase as a result is not known. At best this would be a way of mitigating bad acoustic design.

For open classrooms, systems that allow speech to be directed to specific groups could, if carefully placed and adjusted, provide increased voice level in certain areas while minimising the increase in other areas, potentially reducing intruding noise. This would also require that the space be designed with the system in mind—perhaps similar to the "twenty-first century learning space" described in Mealings et al. (2015b). However the system would need to be very well-controlled.

In either case, the frequency response and other parameters of the electronic transmission chain would need to be optimised as described in Chap. 9, and systems meeting the required specifications are likely to be expensive.

Poorly designed systems, in terms of placement, control or electro-acoustic design, may actually hinder achievement of the acoustic goals described in Sect. 11.1.

11.13 Summary

In a classroom, background sound level, intruding sound and sound quality all need to be directed towards providing a supportive learning environment. This includes providing appropriate STI values, but also includes allowing students and teachers to feel comfortable. Several results reported above emphasise this point:

- background levels above 40–45 dBA L_{Aeq} result in noisier classes and greater "total disruption time" (Shield et al. 2015),
- speech perception depends on ambient noise level and distance from the speaker, but also on other properties of the room (Mealings et al. 2015b),
- children's task performance in noise depends more on the type of noise than on its level (Dockrell and Shield 2006),

- increased reverberation results in more "dysfunctional activities" and greater teacher stress (Tiesler et al. 2018).

With this in mind, the following recommendations are offered for classroom design (although as always specific circumstances need to be considered):

1. Keep $RT_{30} < 0.5s$, 500 Hz–2 kHz. Because sound quality appears to be crucial, also keep late reverberation below, say, 0.6s, and $C_{50} > 6$ dB.
2. At lower frequencies of 125 Hz and 250 Hz, keep $RT_{30} < 0.6s$ to ensure that speech intelligibility is not compromised by "upward masking" as described in Chap. 10.
3. For open classrooms, use barriers and separation as far as possible. The aim is STI > 0.5 for all students in a class group with noisy activity in the adjacent area.
4. For open classrooms, put high-quality absorption ($\alpha > 0.9$ @ 1 kHz) on the ceiling to minimise propagation over barriers and separators.
5. For enclosed classrooms, partitions between classrooms should have $D_{nT,w} \geq 45$.
6. Background SPL (unoccupied) should be less than 40 dBA L_{Aeq}.

References

ANSI. ANSI S12.60/Part1–2010 acoustical performance criteria, design requirements, and guidelines for schools, Part 1: permanent schools

Association of Australasian acoustical consultants. Guideline for healthcare facilities V2.0

Association of Australasian acoustical consultants. Guideline for educational facilities, V2.0

Christensson J (2018) Good acoustics for teaching and learning. In: Proceedings of Euronoise 2018

Dockrell J, Shield B (2006) Acoustical barriers in classrooms: the impact of noise on performance in the classroom. Br Educ Res J 32(3):509–525

European Standards (2016) DIN 18041: acoustic quality in rooms—specifications and instructions for the room acoustic design

Harvie-Clark J (2014) Use of G and C50 for classroom design. Proc Inst Acoust 35(3)

Knect H, Nelson P, Whitelaw G, Feth L (2002) Background noise levels and reverberation times in unoccupied classrooms: prediction and measurement. Am J Audiol 11:65–71

Mealings K, Buchholz J, Demuth K, Dillon H (2015a) Investigating the acoustics of a sample of open plan and enclosed kindergarten classrooms in Australia. Appl Acoust 100:95–105

Mealings K, Demuth K, Buchholtz J, Dillon H (2015b) The effect of different open plan and enclosed classroom acoustic conditions on speech perception in Kindergarten children. JASA 138(4):2458–2469

Morales K, Li F (2018) A new verification of the speech transmission index for the English language. Speech Commun 105:1–11

Sato H, Bradley J (2008) Evaluation of acoustical conditions for speech communication in working elementary school clasrooms. J Acoust Soc Am 123(4):2064–2077

Shield B, Dockrell J (2004) External and internal noise surveys of London primary schools. JASA 115(2):730–738

Shield B, Conetta R, Dockrell J, Connolly D, Cox T, Mydlarz C (2015) A survey of acoustic conditions and noise levels in secondary school classrooms in England. JASA 137(7):177–188

Standards Australia (2016) Standards New Zealand, AS/NZS 2107:2016 Acoustics—recommended design sound levels and reverberation times for building interiors

Stansfeld S, Berglund B, Clark C, Lopez-Barrio I, Fischer P, Ohrstrom E, Haines M, Head J, Hygge S, van Kamp I, Berry B (2005) Aircraft and road traffic noise and children's cognition and health: a cross-national study. The Lancet 365:1942–1949
Tiesler G, Machner R, Brockmann H (2018) Classroom acoustics and impact on health and social behaviour. Energy Procedia 78:3108–3113
U.K. Department of Education (2015) Building bulletin 93. Acoustic design of schools: performance standards
Valentine J, Wilson O (2002) Classroom acoustics: a New Zealand perspective, [Online]. Available: https://www.oticon.org.nz/pdf/classroomacousticsreportprintedversion.pdf
Whitlock J (2012) Understanding the Lombard effect. N.Z. Acoust 25(2):14–17
Wilson W, Armstrong R, Perrykkad K, Arnott W (2019) The acoustic health of primary school classrooms in Brisbane, Australia. Speech, Lang Hear 23:1–8

Chapter 12
Hospital Wards

Abstract This chapter follows a similar pattern to Chap. 11, describing acoustic design principles for hospital wards. Once again, criteria and recommendations for background sound levels, intruding sound, reverberation and speech intelligibility are discussed, in this case including the potential for sleep disturbance. Measured values of acoustic parameters are reported, and the results of a number of interventions are discussed in detail. Once again, the chapter concludes with a summary of best practice for the acoustic design of these spaces.

12.1 Approach to Acoustic Design

Health-related spaces include (among others)

- hospital wards—ICU and general,
- hospital nursing stations,
- consulting rooms,
- gyms and exercise areas,
- audiometric testing facilities.

As for teaching spaces, these all have specific, and different, acoustic requirements. This section will concentrate on hospital wards, and ICU wards in particular, which have the most unique acoustic requirements.

As for teaching spaces (and all other spaces), the major acoustic elements are:

- **Background Sound** (that is, constant sound when the space is unoccupied),
- **Intruding Sound** (that is, sound from speech, monitors, alarms, etc., in the same room, or from other spaces),
- **Sound Quality** (RT and other indicators of sound quality),
- **Speech Intelligibility** (including speech privacy).

Vibration control may also be an issue where vibration-sensitive equipment is used.

R. Bullen, *The Architecture of Sound*, Springer-AAS Acoustics Series,
https://doi.org/10.1007/978-981-97-9380-8_12

Table 12.1 Recommended background sound levels in hospital wards

Document	Description of space	Recommended background SPL
AS/NZS 2107:2016 (Standards Australia 2016)	Ward bedrooms	Range 35–40 dBA L_{Aeq}
	Intensive care wards	Range 40–45 dBA L_{Aeq}
	Nurses' stations	Range 40–45 dBA L_{Aeq}
AAAC Guideline for Healthcare Facilities (Association of Australasian Acoustical Consultants)	Single bed ward	40 dBA L_{Aeq} day 35 dBA L_{Aeq} night 50 dBA L_{ASmax}
	Multiple bed ward	40 dBA L_{Aeq} any time 50 dBA L_{ASmax}
U.K. Health Memorandum HTM 08–01 (U.K. Department of Health 2013)	Wards	40 dBA L_{Aeq} day 35 dBA L_{Aeq} night
USA—FGI Guidelines (Institute 2010)	Patient rooms	≤35 dBA $L_{Aeq,1\,h}$
WHO (World Health Organisation 2009)	Sleep in general*	35 dBA L_{Aeq} 40 dBA L_{AFmax}

*The wording in the document is confusing. Most people take the recommendations to refer to ambient noise, in which case they are completely unachievable. If they refer to background noise, they are roughly consistent with other recommendations

12.2 Background Sound in Hospital Wards—Recommendations

Table 12.1 shows guidance from various documents as to appropriate background sound levels in hospital wards.

12.3 Background Sound in Hospital Wards—Measured

Measured background levels in hospital wards are not often reported—most reports are of ambient noise. One report for Johns Hopkins Hospital, Baltimore (Busch-Vishniac et al. 2005), gives noise levels in unoccupied wards of

38–43 dBA L_{Aeq} daytime,
37–38 dBA L_{Aeq} nighttime.

A report for "a Sydney hospital" (Elliott et al. 2010) gives L_{A90} levels in an occupied ward of 44–50 dBA, which may approximate background levels.

In general, background levels in wards appear to be just within or somewhat above recommendations. They are generally from mechanical services and/or equipment in the ward and can be controlled by standard methods. However, intruding ambient levels are much higher and are generally of greater concern.

12.4 Intruding Sound in Hospital Wards—Recommendations

In a hospital ward, intruding sound may come from outside the ward, but in an ICU or multibed ward, it is most likely to come from equipment or activities within the same ward. The most significant potential effects of this noise are:

- patient discomfort/stress,
- staff discomfort/stress,
- sleep disturbance.

Criteria in the AAAC guidelines and UK Health Memorandum appear to be directed to noise intrusion from other rooms. They are:

AAAC Guideline for Healthcare Facilities (Association of Australasian Acoustical Consultants):

$L_{\text{ASmax}} < 50$ dBA.

U.K. Health Memorandum HTM 08–01 (U.K. Department of Health 2013):

$L_{\text{AFmax}} < 45$ dBA.

As described below, when applied to the total noise environment in a ward, these criteria are probably unachievable.

12.5 Measured Sound Levels in Hospital Wards

There are numerous reports of recorded sound levels in hospital wards, expressed in various noise units. Table 12.2 shows measured levels from individual events in an ICU ward in a Sydney hospital, expressed as L_{AFmax}, from Elliott et al. (2010).

Figure 12.1 shows reported noise levels in various wards at Johns Hopkins Hospital, expressed as logarithmic averages of what appear to be 30-min recorded values. It is interesting to note in passing that "acoustic ceiling tiles" appear to reduce noise levels by about 3–4 dB.

Figure 12.2 shows reported night-time noise levels in a tertiary hospital in Germany, expressed as what appear to be 1-min L_{Aeq} values.

Table 12.2 Recorded noise levels from individual events in an ICU ward, from Elliott et al. (2010)

Noise source	Measured L_{AFMax}, dB
Conversation	61
Dressing trolley moved	77
Intravenous fluid pump alarm	73
Mattress alarm	67
Monitor alarm	62

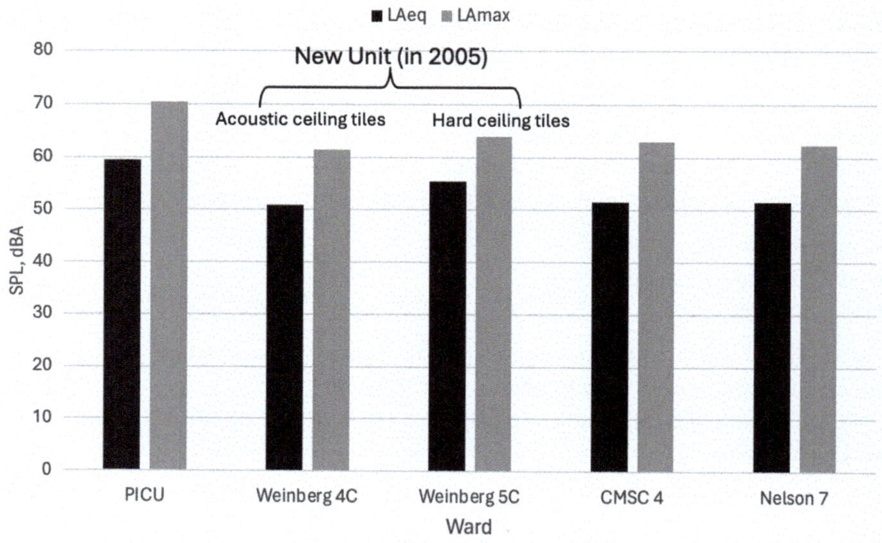

Fig. 12.1 Recorded noise levels in various wards at Johns Hopkins Hospital, adapted from Busch-Vishniac et al. (2005)

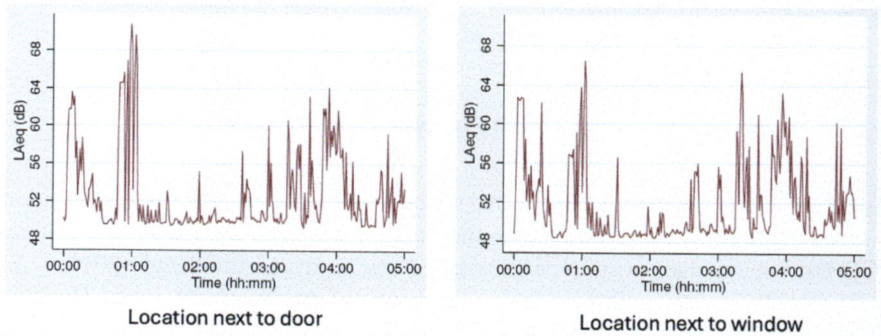

Fig. 12.2 Recorded noise levels in one ward at a German tertiary hospital, before treatments, from Leutz et al. (2016)

Despite the disparity in noise units, it is clear that maximum noise levels in hospital wards throughout the world are consistently well above the criteria in Sect. 12.4. The potential impact of these levels, particularly on sleep, is discussed in Sect. 12.7.

Table 12.3 Recommended reverberation times or treatments in hospital wards

Document	Description of space	Recommended reverberation time/ treatment
AS/NZS 2107:2016 (Standards Australia 2016)	Ward bedrooms	0.4–0.7 s
	Intensive care wards	<0.6 s
	Nurses' stations	0.4–0.7 s
AAAC guideline for healthcare facilities (Association of Australasian Acoustical Consultantss)	Single bed ward	Ceiling equivalent to sound absorption Class B under AS ISO 11654
	Intensive care ward	Ceiling equivalent to sound absorption Class A under AS ISO 11654
U.K. health memorandum HTM 08–01 (U.K. Department of Health 2013)	Wards	"Sound absorbent treatment should be provided in all areas ... except acoustically unimportant rooms ..."

12.6 Reverberation Times in Hospital Wards—Recommendations

Recommended reverberation times are provided in various documents as shown in Table 12.3.

There are very few reports of measured reverberation times in hospital wards. Some are discussed below under "Interventions".

12.7 Impacts on Sleep

The effect of noise events on sleep has been extensively studied in the context of the effect of environmental noise on the sleep of residents. Studies of single events, such as aircraft passbys indicate that depending on their maximum level, they may be associated with motility (body movements recorded by an accelerometer) and changes in sleep state (recorded by an electroencephalogram). One conservative definition of "awakening" is a change to either the "Awakened" or "Stage 1" sleep state from a deeper sleep state. (These "awakenings" are not necessarily accompanied by waking consciousness and may not be remembered the next day.)

Figure 12.3 shows the relationship between L_{ASmax} of an event and the probability of motility, based on (Miedema et al. 2003), and also a similar result for awakenings defined as above, (Basner et al. 2006). It is notable that the probabilities of motility and "awakening" are similar.

The noise events occurring in a hospital ward do not have the same character as an aircraft passby and the background level from which they emerge is also different. Nevertheless, these results provide some context for the noise levels reported in Sect. 12.5. At a level of 60 dBA L_{ASmax}, which appears to be common in wards, there

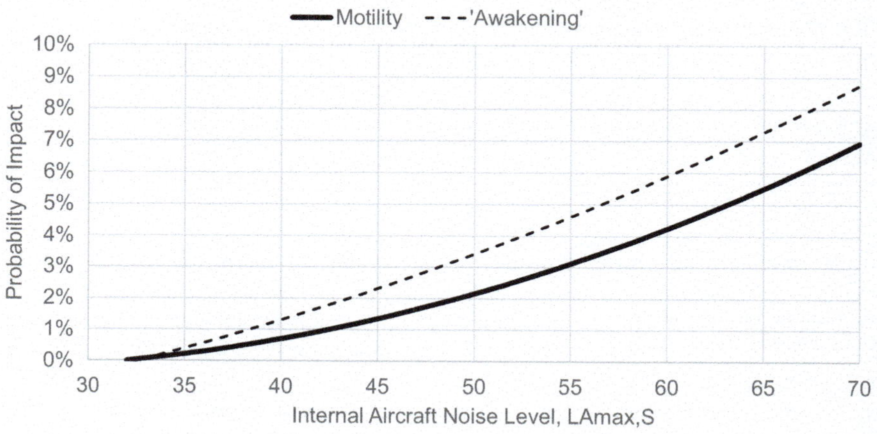

Fig. 12.3 Probability of impacts within a 15 s period containing a noise event (aircraft noise)—motility, from Miedema et al. (2003) and "awakening", from Basner et al. (2006)

is about a 5% chance that an aircraft passby would cause motility or "awakening". Given the number of such events per night in wards, these impacts could be expected to occur at least once on most nights for most patients.

Numerous studies have been conducted on the sleep of patients in actual wards. Unfortunately, these vary greatly in their format and methodology. A large review published in 2018 (Horsten et al. 2018) comes to the following disturbing conclusion (emphasis added):

> … there was considerable heterogeneity between studies … and all studies suffered from a considerable risk of bias. The meta-analysis of results was hampered by widely varying definitions of sound parameters between studies and a general lack of detailed description of methods used. **It is, therefore, currently impossible to quantify the extent to which noise contributes to sleep disruption among ICU patients**, and thus, the potential benefit from noise reduction remains unclear. Regardless, the majority of the observed sleep disturbances remain unexplained.

12.8 Interventions

Despite the uncertainty noted above, a number of papers describe interventions that have the potential to significantly reduce noise levels in hospital wards. The interventions are a combination of changes to the building construction and layout and changes to operational procedures. They are described in this section.

12.8.1 Alarms

(Welch 2011) considers ways to reduce the number of false alarms from one specific piece of equipment—a peripheral capillary oxygen saturation (SpO_2) alarm. It finds that by slightly reducing the threshold and increasing the delay before an alarm sounds, the number of false alarms can be reduced by 85% while not affecting the number of true alarms.

12.8.2 Operational Changes

(Li et al. 2011) describes simple interventions to reduce night-time disturbance in an ICU ward:

- close doors,
- dim lights,
- reduce level of telephones and alarms,
- rearrange times for various checks,
- respond to alarms within 1 min,
- reduce level of staff conversations.

With these measures, the reported "peak sound level"—which appears to mean the arithmetic mean of night-time $L_{\text{Amax,1min}}$ levels—reduced from 59 to 51 dBA.

12.8.3 Ward Layout

A comprehensive restructuring of the layout of an intensive care unit ward is described in (Leutz et al. 2016). The changes are shown in Fig. 12.4, where A represents the standard layout and B represents a modified layout. In the modified layout, nurses are stationed in a separate area with large double-glazed windows overlooking the patient area. All traffic (except beds) is through the nurses' area and in many cases does not need to enter the patient area at all. Access to the adjacent patient room is through double doors rather than a single swing door.

Table 12.4 shows measured noise levels in the two cases. While reductions in L_{Aeq} are modest, there was a very significant reduction in the duration of exceedances, particularly those representing individual events. This is reinforced by Fig. 12.5, showing night-time $L_{\text{Aeq,1min}}$ values in the two cases (Table 12.4).

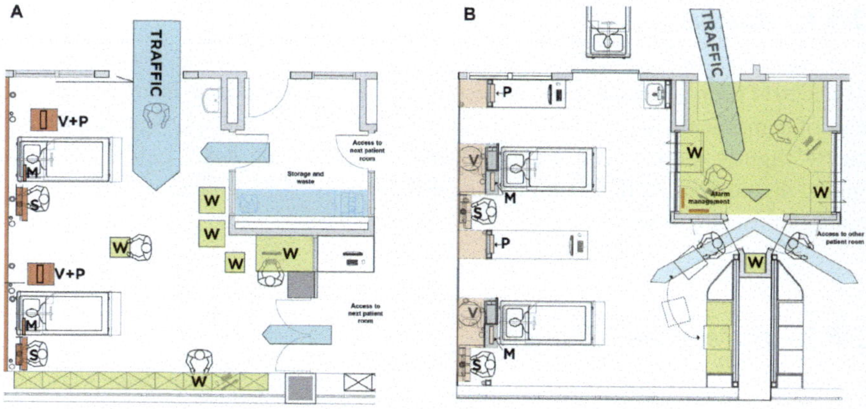

Fig. 12.4 Standard (A) and modified (B) ward layouts—from Leutz et al. (2016). W = workplace; V = ventilator; M = monitor; P = panel (alarms); S = syringe and infusion pumps

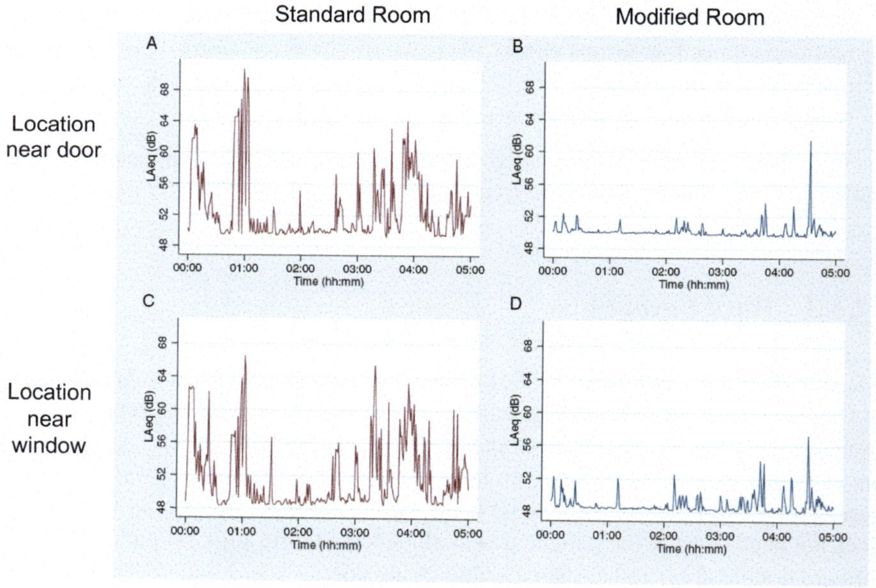

Fig. 12.5 $L_{\text{Aeq,1 min}}$ noise levels in standard and modified rooms, from Leutz et al. (2016)

12.8.4 Reverberation Control

Perhaps most interesting is a study reported in Blomkvist et al. (2005) and (Hagerman et al. 2004). In the coronary care unit of Huddinge University Hospital, patient

Table 12.4 Recorded noise levels in a standard and modified ICU ward, from Leutz et al. (2016)

Noise measure	Standard ward		Modified ward	
	Door	Window	Door	Window
$L_{Aeq,24\,h}$	54	53	52	51
$L_{Aeq,midnight\text{-}5am}$	53	52	50	49
Time $L_{Aeq,1imin} > 50$ (%)	66%	50%	40%	11%
Time $L_{AFmax,1\,min} > 60$ (%)	62%	59%	27%	30%

outcomes were recorded over 20 weekdays with a reflective ceiling and over 22 weekdays after installing absorptive ceiling tiles. The resulting difference in acoustical parameters is shown in Table 12.5.

In terms of patient outcomes, standard clinical measures including heart rate, heart rate variability, blood pressure and overall pulse amplitude showed no significant difference between the two groups. However, three indicators were significantly lower in the low-reverberation conditions:

- **night-time pulse amplitude** for patients with serious heart conditions ($p \sim 0.04$),
- **patient requests for additional beta-blockers** (an indication of pain) ($p < 0.01$): 10% with low RT (6/63) versus 30% with high RT (10/31), and most surprisingly,
- **readmission to hospital after 3 months** ($p < 0.01$): 21% for patients low RT (13/63) versus 48% for patients with high RT (15/31).

The third point is difficult to explain. However, a survey of patients leaving the unit may shed some light on both the second and third points. Although differences between the groups were small for most questions, including for example "disturbed by sounds", there was a large and significant ($p < 0.01$) difference in their rating of "*staff attitude quality*", with the rating being much more positive in the low-reverberation environment.

In other words, the low-reverberation environment was perceived by patients as more caring, and their interactions with staff were more positive, even though this

Table 12.5 Acoustic parameters before and after installing an absorptive ceiling, from Blomkvist et al. (2005)

	Reflective ceiling		Absorptive ceiling	
	Patient room	Central area	Patient room	Central area
RT, s	0.9	0.8	0.4	0.4
$L_{Aeq,24\,h}$	56	57	50	56
STI*	0.67	0.72	0.87	0.88

* These values represent a superseded version of STI called RASTI which tends to produce higher values that the current standard. They are nevertheless useful for comparative purposes

was not perceived as due to a reduction in noise. The acoustic intervention can be seen as soundscape design rather than simply noise control. This could help to explain both a reduced need to request additional pain relief and a better overall clinical outcome.

12.9 Practical Issues

Hospital wards present a number of specific challenges for acoustic control as listed below.

Infection Control

Soft or textured surfaces are generally not permitted in wards. Most surfaces should be washable and resistant to detergents and disinfectants. This includes:

- wall and ceiling finishes,
- floor coverings (carpet is generally not permitted in Australia),
- absorption placed above a ceiling,
- duct linings.

Specialised washable ceiling tiles are available providing reasonable attenuation at mid frequencies, but less than standard tiles at low frequencies.

Mechanical and Electrical Services

There are typically large numbers of ducts, conduits and services passing between spaces in the ceiling void. This makes it difficult to avoid flanking transmission. Once again, specialised washable ceiling tiles are available combining high(ish) transmission loss with high(ish) absorption. However, to achieve the rated values, their installation requires care and control.

Doors

For a wall to achieve a $D_{\mathrm{nT,w}}$ of 45, which would be required for privacy between wards, any door in the wall should achieve at least $D_{\mathrm{nT,w}}$ 40—particularly a large door that facilitates moving a bed. However, doors are often required to swing (in both directions) and often cannot use rubber seals due to infection control. In-cavity sliding doors are preferable for noise control. In any case, design and placement of doors is critical to achieving privacy.

Vibration Control

Some equipment (e.g. pumps) can cause structure-borne noise in other spaces. Control requires standard vibration-isolation techniques.

Some types of hospital equipment (e.g. electron microscopes) have requirements for very low levels of vibration. This requires low-frequency isolators or in some cases complete separation of the equipment from the building structure.

12.10 Summary

With respect to acoustics in hospital wards, and particularly ICU wards, the information above can be summarised as follows.

- Health and acoustic professionals generally agree that hospital wards are too noisy, but good evidence for effects on patients or staff is difficult to find.
- Recommended background SPLs tend to be exceeded, but this may not be a significant issue.
- Intruding noise is significant and there is evidence that it can cause discomfort and sleep disturbance as well as stress for staff.
- The most effective ways to control intruding noise are through design layout and staff protocols, particularly minimising alarms. Reducing RT has some effect on intruding noise levels.
- However, as in classrooms, there is evidence that reducing RT and adjusting other indicators of sound quality have important benefits for the environment apart from reducing SPLs.
- Acoustic mitigation in all spaces needs to take account of special requirements for infection control, mobility and the need for audible alarms.
- There may also be some specialised vibration requirements.

References

Association of Australasian Acoustical Consultants, Guideline for healthcare facilities V2.0

Basner M, Samel A, Isermann U (2006) Aircraft noise effects on sleep: application of the results of a large polysomnographic field study. JASA 119:2772–2784

Blomkvist V, Eriksen C, Theorell T, Ulrich R, Rasmanis G (2005) Acoustics and psychosocial environment in intensive coronary care. [Online]. Available: http://www.occenvmed.com/cgi/content/full/62/3/e1

Busch-Vishniac I, West J, Barnhill C, Hunter T, Orellana D, Chivukula R (2005) Noise levels in Johns Hopkins Hospital. JASA 118:3629–3645

Elliott R, McKinley S, Eager D (2010) A pilot study of sound levels in an Australian adult general intensive care unit. Noise Health 12:26–36

Hagerman I, Rasmanis G, Blomkvist V, Ulrich R, Eriksen C, Theorell T (2004) Influence of intensive coronary care acoustics on the quality of care and physiological state of patients. Int J Cardiol 98:267–270

Horsten S, Reinke L, Absalom A, Tulleken J (2018) Systematic review of the effects of intensive-care-unit noise on sleep of healthy subjects and the critically ill. Br J Anaesth 120(3):443–452

Institute FG (2010) Guidelines for the design and construction of health care facilities

Leutz A, Weiss B, Penzel T, Fietze I, Glos M, Wernecke K, Bluemke B, Dehn A, Willemeit T, Finke A, Spies C (2016) Feasibility of noise reduction by a modification in ICU environment. Psychol Measur 37:1041–1055

Li S-Y, Wang T-J, Vivienne S, Liang S-Y, Tung H-H (2011) Efficacy of controlling night-time noise and activities to improve patients sleep quality in a surgical intensive care unit. J Clin Nurs 20:396–407

Miedema H, Passchier-Vermeer W, Vos H (2003) Elements for a position paper on night-time transportation noise and sleep disturbance, Netherlands organisation for applied scientific research
Standards Australia (2016) Standards New Zealand, AS/NZS 2107:2016 acoustics—recommended design sound levels and reverberation times for building interiors
U.K. Department of Health (2013) Health technical memorandum 08–01: acoustics
Welch J (2011) An evidence-based approach to reduce nuisance alarms and alarm fatigue. Hofrizons Spring:46–52 (2011)
World Health Organisation (2009) Night noise guidelines for Europe

The manufacturer's authorised representative in the EU is Springer Nature Customer Service Centre GmbH, Europaplatz 3, 69115 Heidelberg, Germany. If you have any concerns regarding our products, please contact ProductSafety@springernature.com

Printed and bound by CPI Group (UK) Ltd, Croydon, CR0 4YY
01/10/2026
02225980-0017